MORTAL COIL

DAVID BOYD HAYCOCK
MORTAL COIL
A SHORT HISTORY OF LIVING LONGER

David B. Haycock

YALE UNIVERSITY PRESS
NEW HAVEN AND LONDON

For information about this and other Yale University Press publications, please contact:
U.S. Office: sales.press@yale.edu   www.yalebooks.com
Europe Office: sales@yaleup.co.uk   www.yaleup.co.uk

Set in Adobe Garamond by SX Composing DTP, Rayleigh, Essex
Printed in Great Britain by MPG Books Ltd, Bodmin, Cornwall

Library of Congress Cataloging-in-Publication Data

Haycock, David Boyd, 1968–
    Mortal coil : a short history of living longer / David Boyd Haycock.
        p. cm.
    Includes bibliographical references and index.
    ISBN 978-0-300-11778-3 (alk. paper)
    1. Longevity--History. 2. Immortality (Philosophy) 3. Death.
    I. Title.
    [DNLM: 1. Longevity. 2. Philosophy, Medical--history.
    3. Aging. 4. Health Knowledge, Attitudes, Practice. WT 11.1
    H413m 2008]

    QP85.H39 2008
    571.8'79--dc22
                                                                2007035341

A catalogue record for this book is available from the British Library.

10 9 8 7 6 5 4 3 2 1

Contents

Illustrations

Acknowledgements

When writing my biography of Dr William Stukeley, my interest was caught by a book published in 1722 under the curious title *Long Livers: A Curious History of Such Persons of Both Sexes Who Have Liv'd Several Ages and Grown Young Again*. A few years later, when I was working with Professor George S. Rousseau at De Montfort University, Leicester, he asked me to investigate people who had lived to old age in the eighteenth century for his research into what he playfully called 'the geriatric enlightenment'. Shortly afterwards, I was struck by a remark in a book by Professor Steven Shapin that René Descartes had hoped to live five hundred years.

Profound old age, and the possibility of revived youth, looked like an interesting research proposal. With a reference from Professor Rousseau, in 2003 I was invited to participate as an Ahmanson-Getty Research Fellow at the William Andrews Clark Memorial Library, a satellite of the Center for Seventeenth and Eighteenth Century Studies at the University of California, Los Angeles. Professor Max Novak was running a nine-month programme, 'The age of projects: Changing and improving the arts, literature and life during the long eighteenth century, 1660–1820'. It was there that much of the research for this book was undertaken, and I am therefore particularly grateful to Professors Rousseau and Novak, and to the four other Fellows who participated in the programme: Dr Martin Gierl, Dr Sarah Kareem, Dr Kimberley Latta and Dr Alison O'Byrne. The hospitality of everyone at the Clark Library was fantastic, and I am grateful to all who contributed to my time there, in particular the librarians and the participants in the 'Age of projects' conferences; also to Professor Linda Mitchell for introducing me to Samuel Hartlib and his circle, to Laura Myers for suggesting that this book could be written, and to John Kurtz for providing me with fine accommodation in Los Angeles.

No book on this subject can fail to acknowledge the work of Gerald Gruman. His substantial 1966 essay beat the path that I have followed. This book has been completed in spaces between my most recent project, a Wellcome Trust Research Fellowship working with Dr Patrick Wallis in the Department of Economic History at the London School of Economics and Political Science. Some of this work, on early modern medicines, has fed into my first two chapters; I have also been able to continue my interest in ageing through Professor Mary Morgan's kind invitation to participate in the department's Leverhulme Trust-funded project, 'The nature of evidence: how well do "facts" travel?' The lively discussions of this group have been a real pleasure, and I am particularly thankful to Mary and Patrick, and to everyone else in the 'Facts' team, for their contribution.

A number of people have kindly read individual chapters or sections of this book at various stages, and I would like to thank Jonathan Adams, Aubrey de Grey, Ariel Hessayon, Michael Hunter, Lauren Kassell, Jerrold Northrop Moore, Anthony Nanson, Edmund Ramsden, Patrick Wallis, Caroline Warman, Helen Yallop, and my four anonymous referees. Huw Price and Susannah Wilson helped with research for Chapters 3 and 4, and I gratefully acknowledge their assistance and contributions. For taking valuable time out from their research to discuss their work with me, I am very grateful to Professor David Gems of University College London, Dr Aubrey de Grey of the Methuselah Foundation, and Professor John Harris of the University of Manchester. Any errors or misinterpretations are, regrettably, my own.

Many thanks to my agent, Anna Power at Johnson & Alcock, for helping to get this book published, to Robert Baldock and Rachael Lonsdale at Yale University Press for seeing it into print, and to Manuela Tecusan for her diligent copy-editing.

As always, none of this would ever have been possible without the constant love and generous support of everyone in my family. Finally I must thank Jonathan Bloom and Susie Wilson for everything: I couldn't hope for better companions.

Preface

'Every Man', Jonathan Swift observed, 'desires to live long; but no Man would be old.'[1] How many, however, have wished to live, youthfully, forever? As a human craving, immortality stretches down the millennia: the greatest surviving monuments of antiquity, the pyramids of Giza in Egypt, were raised (it would appear) in the belief that, beyond this mortal life, there lies another, better, endless existence.

Countless monuments have been erected, cities founded, battles fought, books written, in the hope that our names will not be forgotten, that our deeds will be recorded forever in the annals of history. But what of immortal flesh? What of the desire of escaping death's embrace, and continuing, physically, here on earth – forever?

History abounds with such myths and legends. In the eighth century BC, the Greek poet Hesiod described a distant Golden Age, when people remained perpetually young, living a peaceful, carefree existence in a bountiful paradise. Gilgamesh sought the herb 'Old Man Becomes Young', and discovered it at the bottom of the Persian Gulf. Carrying it home to Assyria, he stopped by a lake to swim; carelessly, he left the plant at the water's edge, where it was eaten by a snake. In Judaeo-Christian tradition, there is the Garden of Eden, where Adam and Eve lived what would have been immortal lives, if only they had not eaten the forbidden fruit.

These are myths. But recent advances in genetic technology and biomedicine seem to promise the opportunity of ever-lengthening youthfulness, ever-longer lives and – perhaps? – one day, immortality itself. As Professor John Harris of Manchester University has recently written in the prestigious journal *Science*: 'New research now allows a glimpse into a world in which aging – and even death – may no longer be inevitable.' It is 'unlikely', Harris adds, 'that we can stop the progression to increased life-spans, and even "immortality"'. He thus advises: 'We should start thinking now about how we can live decently and creatively with the prospect of such lives.'[2]

The potential realization of the dream of perpetual youth may look like a miracle of modern science. But, as I show in this book, it is something that Western scientists and philosophers have been dreaming of – and working towards – for the past four hundred years. Ever since the 'scientific revolution' – the era in which classical and medieval ideas of science, religion and the natural world started to be questioned and rejected – the dream of perpetual youthfulness has tantalised. This book uncovers the archaeology of that profound idea, investigating some of the names (some famous, others almost unknown, but none without interest) who have sought a route to perpetual youthfulness. They include some major names in the establishment of modern Western science, including Sir Francis Bacon, René Descartes and Robert Boyle. In more recent times, there have been Nobel laureates such as the Russian biologist Ilya Ilyich Metchnikoff and the French surgeon Alexis Carrel. Others whose works and interests have crossed the path or influenced this search include Jonathan Swift, Jean-Jacques Rousseau, the Comte de Buffon, Albrecht von Haller, the Marquis de Condorcet, Benjamin Franklin, Christoph Wilhelm Hufeland, William Godwin, Mary Shelley, Thomas Malthus, Charles Darwin, Sigmund Freud, Alexander Graham Bell and Julian and Aldous Huxley.

Not all the people I write about expected, or wanted, immortality: a few decades – or even centuries – more, seemed enough. But underlying this ambition, from Sir Francis Bacon in the sixteenth century through Descartes to such Enlightenment luminaries as Condorcet and Godwin and onto our own day, this was the direction in which these studies clearly pointed. My concern, therefore, is not the immortality of the soul; I leave the history of that immense subject to others. This is the chronicle of the search for the prolongation of physical human life on earth and for the possibility of immortal flesh in Western science and medicine. I chart it from the early years of the seventeenth century right up to the first decade of the twenty-first. It is an immense and, I hope, memorable story.

CHAPTER 1

The History of Life and Death

To live forever, and to become immortall here on earth, is a thing
impossible: but to prolong a mans life free from violent sicknesses, and
to keep the humors of the body in a temperate state, I verily believe it
may be done . . .

Sir William Vaughan, *Directions for Health, Naturall and Artificiall*
(1633)

Late March, 1626 A chill wind blows tufts of snow, seeming reluctant
to touch the icy ground. The philosopher's coach, which he shares with
Dr Witherborne – Scots physician to King James – jogs and rattles
through hardened ruts on the Highgate road. Bound in fur-lined coats,
feet tucked in leather boots, the men are hardly warm. But they are out,
none the less, to take the health-giving country air, free from the
noxious, sulphurous fumes of the London smoke.

Sir Francis Bacon, now sixty-six years old and ailing, takes this daily
journey as a health-benefiting exercise. Though his hair is grey and his
face lined with age, though he has suffered and still suffers those blows
of outrageous fortune which have cast him down from highest political
office, he is today as much filled with ideas and exuberance and
imagination as he was more than forty years before. Then, as a young
student at Cambridge University, he had first questioned the teachings
of Aristotle. And now a thought comes to his lively mind. He has often
wondered if flesh, like fruit, might not be long preserved in snow, and

quite as well as it is in salt. With his vivid, viper eyes, he has spied something from the window of the coach, and here is a chance to conduct an experiment. He calls loudly for his huddled driver to halt. As the coach horses blow air like steam through their nostrils, the philosopher and the doctor clamber from the carriage, trudge through snow, and knock at the door of a cottage at the foot of Highgate Hill. A poor woman opens it, no doubt amazed to see such eminent visitors. The philosopher produces coins and buys the chicken he had seen picking at scraps in the yard. The woman kills and eviscerates it, and the men help to stuff the bloody carcass with snow, before returning to their coach with the frozen bird. But the cold has chilled the old man, who has long been ill and weakening. His fingers swell red and stiff. He is suddenly overcome with fever; he falls into a fit, vomiting.

Dr Witherborne, realizing that Sir Francis is too weak to return to his Holborn lodgings, directs the coachman to the nearby home of Thomas Howard, Earl of Arundel. The Earl and his wife, however, are absent – imprisoned in the Tower of London on a whim of the king – and the beautiful Italian gardens of his Highgate villa are lost beneath sheets of snow. But the earl's servants prepare a guest-room for their eminent ailing visitor, warming the long-empty bed with a heated pan.

From there, on the following day, Sir Francis dictates a letter to his secretary, who has ridden up from London (perhaps bringing with him one of his master's medicines: opium and rose water, perfumed with cloves). He requests 'humble thanks' from the earl 'for a favour' and apologises for this unexpected stay. He had, he explains, been 'desirous to try an experiment or two, touching the conservation and induration of bodies. As for the experiment itself, it succeeded excellently well . . . But when I came to your Lordship's house, I was not able to go back, and therefore was forced to take up my lodging here, where your house-keeper is very careful and diligent about me.' He apologises for not having written the letter personally, 'but in troth my fingers are so disjointed with this fit of sickness, that I cannot steadily hold a pen'.[1]

Over the next few days the philosopher's cold and fever worsen. The bed, unslept in for a year, is damp. What was at first a cold advances into bronchitis – a deadly inflammation of the mucous membrane of the lungs. On the morning of Easter Sunday, the old man chokes on a discharge of bodily fluids, so copious it drowns him. Sir Francis Bacon,

Lord Verulam and Viscount St Albans, the greatest English philosopher of his age, is – somewhat unexpectedly – dead.[2]

§

Why, exactly, did Sir Francis Bacon stop his coach that freezing winter's day to stuff a chicken with snow? The answer is probably this: ironically, his fatal, icy experiment was part of his dream of restoring humans to that immense longevity they had enjoyed before the Fall. 'Knowledge', as Bacon famously declared, 'is power.' And 'the true

Figure 1.1 Sir Francis Bacon: engraving by Wenceslaus Hollar (after a design probably by John Evelyn), from the frontispiece to Thomas Sprat's *History of the Royal Society of London* (1667). In the *History of Life and Death* (1623), Sir Francis Bacon laid out his ideas on how he believed human immortality might be achieved. In this allegorically charged representation, the long-dead Bacon is on the right, sitting with the Royal Society's first president, William Brounker, beside a bust of its founding patron, Charles II.

ends of knowledge', as he observed in 1603, involved the restoration of humankind to 'the sovereignty and power' over the whole of nature. This was the knowledge and authority Adam had briefly enjoyed in the Garden of Eden, when he had given names to all the animals in creation. But it was not only this: 'to speak plainly and clearly', wrote Bacon, 'it is a discovery of all operations and possibilities of operations from immortality (if it were possible) to the meanest mechanical practice'.[3]

Who was this philosopher who thought it might be possible to live young and healthy, perhaps forever? Francis Bacon was born in London in 1561, second son of Sir Nicholas Bacon, a self-made man who had risen from the ranks of Suffolk yeomanry to become Queen Elizabeth's Lord Keeper of the Privy Seal. Francis grew up at his father's country estate at Gorhambury, near St Albans in Hertfordshire. At the age of twelve he went to Trinity College, Cambridge; at fifteen he travelled to France as companion to the English ambassador; at seventeen he began legal studies at the Inns of Court in London. From there, his rise to power was slow but steady: in 1594 Elizabeth made him one of her learned counsel; under James I he became attorney general; then, in 1618, he achieved the highest legal position in the land, that of Lord Chancellor. In his brief years of power he kept a well-heeled, well-dressed court that was said to rival the king's. He was arrogant, ambitious, ostentatious, homosexual, and it was alleged that Bacon's ganymedes and favourites took bribes on his behalf.[4] The poet Alexander Pope would later characterize him as 'The wisest, brightest, meanest of mankind'.[5] When it came, his fall was sudden and rapid: in 1621 he was accused of accepting expensive gifts from plaintiffs, was impeached, and confessed 'I am guilty of corruption'.[6] He was amazed, since his crime was not uncommon.

He was fined, briefly imprisoned in the Tower of London, and banished from court. His faithless friends and handsome servants vanished. He sold his beloved house in the Strand, retiring 'from the stage of civil action' to his private chambers at Gray's Inn. There he turned his full attention during his final years to the reform of philosophy and science.

Once Bacon was released from the affairs of the English Renaissance state, his plan was to write six essays which would form the third part of his 'Great Instauration', the comprehensive philosophical programme he had been preparing and part-publishing over the past twenty years. The first of these essays would be on winds, the last, on life and death. But, on finishing his *Historia Ventorum*, he changed his mind about their order. Given what, he explained, was 'the extreme profit and importance of the subject, wherein even the slightest loss of time should be accounted precious',[7] why wait till he had written four more essays? There could be no hanging around in the search for immortality!

Thus, over the winter of 1622, Bacon wrote his newly promoted dissertation – he called it *The History Naturall and Experimentall, of Life and Death*. It was completed in Latin in January 1623 and published later that year. It carried the subtitle 'of the Prolonging of Life'.[8] In Bacon's youth the French physician Laurent Joubert, Chancellor of the University of Montpellier, had pondered the question whether it was possible for medicine considerably to prolong the life of men, observing that such speculation 'has always been intense and has excited the greatest minds'. Objectively reviewing both sides of this argument in the second chapter of his best-selling *Erreurs populaires au fait de la médecine et régime de santé*, Joubert had concluded that it *was* possible to 'elongate the terms of all ages, and thus of all life, by medicine, even further than is ordered by Nature'.[9]

But Joubert had not proposed a clear way this was to be done. Bacon's *History of Life and Death*, in the depth of its detail and the range of its applications and speculations, was a work unprecedented. Bacon was well qualified to undertake the task: as his secretary and executor William Rawley subsequently observed, 'if there were a Beame of Knowledge, derived from God, upon any Man, in these Modern Times, it was upon Him'.[10] Even a Frenchman, writing in the 1660s, could remark that Bacon 'undoubtedly is the greatest Man for the Interest of Natural Philosophy that ever was'.[11]

§

The fundamental framework for Bacon's belief that life could be vastly prolonged was the Bible. The first chapters of Genesis make it clear that, if Adam and Eve had not eaten the forbidden fruit, they would have enjoyed eternity on earth. In the Middle Ages the Oxford friar Roger Bacon had recorded that 'man naturally is immortal'.[12] In 1644 the puritan pamphleteer Richard Overton began his tract, *Man's Mortalitie*, by explaining that God created Adam and Eve 'both innocent and free from sin, and so from *Death* and mortality: *For the wages of Sin is Death* . . . Thus Man was gloriously immortall.'[13] It was in Eden that mankind had lost this golden opportunity of illimitable life – in fact, had been denied that authority over nature that God had first granted to Adam and to which Francis Bacon hoped we would one day return.

Sin had frayed the sturdy fabric of God's creation. 'If you inquire therefore into the ruines of human nature,' wrote Richard Steele in 1688, 'the answer will be, that *Sin* is the *moth*, which, being bred therein, hath fretted the garment, withers the man, and layes his honour in the dust.'[14] We were pale imitations of our first parents: in Bacon's day (a time when those who cared about such things believed God had created earth on an autumn afternoon in 4004 BC), it was held that Adam had possessed universal, encyclopaedic knowledge and a perfect memory. According to the philosopher Joseph Glanvill, writing in 1661, Adam's understanding of the world's complexities had been enhanced by the power of telescopic sight, so that he could see in all its finery 'the Coelestial magnificence' of the heavens. Perhaps, Glanvill thought, Adam had even possessed microscopic sight, so that 'he saw the motion of the bloud and spirits through the transparent skin'.[15]

Not only were Adam and Eve perfect in knowledge, they were physically faultless, too. According to the late Elizabethan astrologer Simon Forman, 'when he was created ther was noe creatur in beauty shape and wisdom like Adam'. His genitals had not embarrassed him, because 'before his Fall he had no genitors, but [only] after he was put forth of paradice, his genitors began to growe forth of him'.[16] According to George Walker, writing in 1641, Adam and Eve's skin was 'faire, white, and ruddie, was comely in itselfe, and beautifull to their own eyes'. For Samuel Pordage, writing twenty years later, Adam had been gifted with incredible strength, and was physically indestructible.[17] There was seemingly nothing that this almost bionic, immortal man had not seen or understood. It was obvious to anyone in seventeenth-century England that modern men and women fell far short of such delightful excellence.

It was, of course, the Fall that cut short this all-too-brief Golden Age. God had told his first children that they could eat the fruit of any of the trees in Eden, *except* that from the one growing at its centre. 'God told us not to eat the fruit of that tree or even touch it', Eve said to the Serpent: 'if we do, we will die.' But this was the Tree of the Knowledge, the Serpent explained. He encouraged her to eat, and Eve, thinking 'how wonderful it would be to become wise', ate the fatal fruit. And Adam also ate the fruit. Then, realizing they were naked, in shame they covered their genitals, and hid themselves from God.[18]

But no one can hide from God. Discovering that they had eaten from the Tree of Knowledge, He cast them from Eden. 'Then the Lord God said, "now the man has become like one of us and has knowledge of what is good and what is bad. He must not be allowed to take fruit from the tree that gives life, eat it, and live forever." So the Lord God sent him out of the Garden of Eden and made him cultivate the soil from which he had been formed.'[19] In the opinion of Simon Forman, the forbidden fruit had acted like a poison on the bodies of our first ancestors. Adam 'becam monstrous and lost his first form and shape divine and heavenly and becam earthy full of sores and sickness for evermore'.[20] According to the Cambridge apothecary Robert Talbor, writing in 1672, since the Fall both 'Soul and Body' of men 'have deviated from the first perfection . . . the Memory is subject to fail, the Judgement given to erre, and the Will often known to rebel, and become a voluntary slave to passion: so is his Body subject to so many infirmities'.[21] It was from this moment on that sickness entered the human frame: 'Take away the Curse, and the forbidden Fruit,' wrote the German chemist Albert Otto Faber in 1677, 'and all Diseases will vanish, and Life become Free again.'[22]

The Fall, however, could not be undone. With the sins of the father passing on to the son there began the era of long and steady human decline, down to Bacon's age. In seventeenth-century Europe, this was the official history of mankind. It was a brave soul who dared suggest otherwise.[23]

Yet this was not the whole story. Though Adam and Eve had forfeited the gift of immortality, they and their offspring still lived many hundreds of years. According to Genesis, Adam lived 930 years, and Noah 950. Methuselah, Noah's grandfather and the eldest of all the patriarchs, died at the ripe old age of 969. The stoicizing Neoplatonist Flavius Josephus (in his *Jewish Antiquities*) and the classically trained Church Father St Augustine (in his *City of God*) defended the literal interpretation of these ages.[24] Given such authoritative champions, the figures were not widely questioned in Bacon's day.[25] Indeed, in 1619 William Basse pondered why God had not let Adam and Methuselah live to the round figure of a thousand years. 'This', he wrote, 'is not without some deep mystery.' He thought the answer 'may be partly because a 1000 yeares hath a type of perfection, [and] God never suffered any to fulfill it, to shew that there is no absolute perfection in this world'.[26]

In 1646, the literary physician Sir Thomas Browne questioned the supposed ages of some biblical characters, but insisted that 'of those ten mentioned in Scripture with their severall ages it must be true'. Indeed, he pointed out that other patriarchs, whose lifespans were not given in the sacred text, might actually have lived *longer* than Methuselah. He rejected the suggestion made by some critics that these so-called years had in fact been lunar months – a calculation that would have made Methuselah a more plausible 90-odd when he died. This theory was wrong on a number of counts, Browne explained, not least because it ran into 'an absurdity, for they make Enoch to beget children about six years of age; for whereas it is said he begat Methuselah at 65'.[27]

It was clear, too, from the Bible that these patriarchal lifespans had steadily fallen. In 1670 the London physician Edward Maynwaringe observed that in 'the *Primitive* Age of the *World*, mans life was accounted to be about 1000 *Years*: but after the *Flood*, the Life of Man was *abbreviated* half'. It had continued to fall, such that by the time of Moses it was 'commonly not exceeding 120 Years'. And '[n]ow the *Age* of Man is reduced to half that: 60 or 70 years we count upon'.[28] This was the famous lament in the Book of Psalms: 'The days of our years are threescore years and ten; and if by reason of strength they be fourscore years, yet is their strength labour and sorrow; for it is soon cut off, and we fly away.'[29]

But was that to be it? Was seventy or at most eighty years the best we could now expect? Were we to content ourselves, instead, with only an immortal soul? For some, the answer was yes; to wish otherwise was to mark oneself as impious, afraid of God's final judgement. As the Scots Calvinist preacher Zacharie Boyd declaimed in a sermon of the 1620s: 'Though the house were never so strong, at last it must decay and drop thorow. There is no lodging for eternitie in things below. Methuselah with his nine hundred three score and nine yeeres is followed with, *hee died*, as well as hee who lived but an houre.'[30] Whatever its length, in the face of heavenly eternity human life was short, and death inevitable. Only the atheist trembled in the face of it.

For Bacon (as it has been for many other philosophers, scientists and physicians since), this was not enough. He did not consider impious the search for the prolongation of life. Rather, it reflected a growing confidence that man could regain his lost control over nature. The physical – surely – could be restored, renewed and *improved*, till once more we lived long and knew all, as had our ancestor Adam in Eden.

§

It is a common impression that life in early modern England was, to misappropriate Thomas Hobbes's famous phrase, 'nasty, brutish and short'.[31] This impression is false. Certainly many, many people died in infancy, and this reduced early modern life expectancy at birth to a frighteningly low figure. Illness was widespread at all ages, and accidents common: London's weekly Bills of Mortality listed over 150 diseases and other different ways of dying. Fevers, consumption and smallpox were among the leading endemic causes of death in northern Europe, whilst epidemics such as plague could cut large swathes through populations.[32]

But those who survived could sometimes live long. It has been calculated that, in Stratford-upon-Avon in the period 1570 to 1630, almost a third of the adult men and a fifth of the women lived beyond sixty years.[33] The average age of the nine seventeenth-century Archbishops of Canterbury was sixty at appointment, and their average age at death seventy-three.[34] In New England, half of the men and women who survived beyond twenty could expect to live into their sixties, whilst for the whole population of Old England at the same time a man of twenty-five had a one-in-sixteen chance of reaching eighty.[35] In 1695 Gregory King estimated that one in ten of the population of England was sixty or over, whilst in October 1702 the writer John Evelyn celebrated his eighty-second birthday: he gave thanks to God that he had retained his intellect and senses 'in greate measure above most of my Great Age'.[36] Some solid few (including Thomas Hobbes, who died in 1679, the Baptist minister Hanserd Knollys, who died in 1691, and the architect Christopher Wren, who died in 1723) lived on into their nineties.[37] William Badger, Member of Parliament for Winchester in 1597, was supposedly born around 1523 and buried in January 1629. It has been suggested that 'there is a fairly solid case' that Badger 'died a centenarian', but this seems unlikely.[38] Sir John Holland, another MP, was certainly born in Norfolk in October 1603 and died on 19 January 1701; at ninety-seven he appears to have been the longest-lived Englishman of this era whose dates can confidently be verified. The Oxford graduate and defrocked clergyman John Humfrey soon surpassed even this record: he was baptized in January 1621 and died in 1719.[39]

Bacon was not mistaken, therefore, in believing it *was* quite possible to live well over seventy years. Though he rejected what he considered the unreliable examples of longevity recorded in 'heathen authors', he filled many pages of his *History of Life and Death* with names of people, in both ancient and modern times, who had lived beyond eighty. In fact, he reckoned that there was 'scarce a *Village*' in England 'but it affords some Man or Woman of Fourescore yeares of Age', noting that he had himself once met in court 'an *old Man*, above an hundred yeares of Age'.[40] He also recorded that it was 'reported' that an Irish contemporary, the Countess of Desmond, had 'lived to an hundred and forty yeares', whilst the inhabitants of the Barbary mountains in north Africa, 'even at this day, they live, many times, to an Hundred and fifty yeares'.[41] Later in the century Sir William Temple – English Ambassador to The Hague and a scholarly man – wrote that the native Brazilians were said 'to have lived two hundred, some three hundred Years'.[42]

In an era when, as Mary Abbott observes, old age 'was much less clearly defined' and birth certificates nonexistent, contemporary accounts would seem to corroborate Bacon's notion that great longevity remained possible.[43] Numerous examples can be given. The Northampton doctor James Hart wrote in 1633 that to 'attaine to 100 is no wonder, having my selfe knowne some of both sexes'.[44] The compulsive recorder of London obituaries, Richard Smith, noted the deaths of two men in 1667, both aged 'about 100 years', and of an old woman in 1670 'above 100'. In Oxford, the antiquary Anthony Wood noted the passing of two local women, both aged 104, in 1679 and 1680;[45] and in 1681, in the same city, the physician and philosopher John Locke recorded a long conversation with an old woman named Alice George. She told him that she was 108, and that her grandmother had lived to 111: 'her hearing is very good,' Locke observed, 'and her smelling so quick, that as soon as she came near me, she said I smelt very sweet, I having a pair of new gloves on that were not strong scented.'[46] Sir Thomas Browne recorded a curious case of what he called *boulimia centenaria*: a tiny old woman 'now living in Yarmouth named Elizabeth Mitchell, an hundred and two years old', who 'greedily' drank and ate as much as she could, 'day and night'.[47] The Oxford natural historian Robert Plot remarked in his *History of Staffordshire* (1686) that James Sands had died in that county in 1588, aged 140.[48] And a few years after Plot's account, Sir William Temple wrote of meeting a beggar at a Staffordshire inn who professed to be 124; this old man told Temple that he ate milk, bread and cheese, and meat only rarely, and drank mostly water.[49] Finally (though this is hardly the last example I could cite) the papers of Robert Boyle include a recipe for a 'Medicine for clearing of the eye-sight found out by Dr. Purlow Sometime Bishop of Hull and Suffragan of York who at the age of 125 years was able to read any Print without Spectacles which att the age of 50 he could no'.[50]

Nonetheless, centenarians were considered unusual. Richard Steele reckoned that 'an *hundred thousand are dead* and rotten, for [every] one that reach such Longevity'. Steele, who was only fifty-eight when his *Discourse Concerning Old-Age* was published in 1688, already placed himself among 'the Weaker sort of Ancient persons'; in 1648, at seventy-five, the architect Inigo Jones described himself as 'being then very old'.[51] Furthermore, it is doubtful that many (or quite probably *any*) of these abovementioned centenarians were as old as claimed. If Boyle's 'Dr. Purlow' is the Robert Pursglove, suffragen bishop of Hull and prebendry of York, who died in 1580, then he was only in his mid-seventies when he died; it is possible that Alice George, who told Locke she was 108, was actually eighty-five. As Thomas Fuller sagaciously observed in 1647, 'many old men use to set the clock of their age too fast when once past seventie; and growing ten yeares in a twelve-moneth, are presently fourscore, yea, within a yeare or two after, climbe up to an hundred'.[52] Exaggerating their age was one way for the old to draw attention to themselves; for, as Keith Thomas has suggested, they were becoming increasingly marginalized in this period – especially if they were poor.[53]

In terms of modern research into seventeenth-century geriatrics, this is important.[54] But what really matters here is that the recorders of all these abovementioned examples apparently *believed* their subjects were as old as they claimed. Indeed, it was widely taken as hard fact in the seventeenth century that the natives of certain parts of America lived up to *three hundred* years.[55] And there was further evidence to support the case for extreme contemporary longevity. The most famous example was Thomas Parr. Coincidentally (or perhaps not), it was the Earl of Arundel, in whose damp bed Francis Bacon had died, who in 1635 discovered the supposedly 152-year-old Parr, blind but living a healthy, humble married life in rural Shropshire. Parr had married his much younger second wife, he claimed, at the age of 112. At a time when it was widely assumed that sexual activity ended at around sixty-five, it was reported that Parr 'had had intercourse' with his new spouse 'exactly as other husbands do, and had kept up the practice to within twelve years of his death'.[56]

Modern research has unearthed a document from 1588 confirming that Parr was a married man in that year; assuming that this was his first wife and that he had married when aged about twenty, he would have been at least sixty-seven in 1635.[57] There were sceptics at the time, who pointed out that there was no way of proving Parr's age. As John

Figure 1.2 Thomas Parr: engraving after an unknown artist, published in 1821 by T. and H. Rodd. A plaque still marks the place in Westminster Abbey where Thomas Parr was buried in 1635, supposedly at the age of 152. His achievement entered legend: his image was reproduced in popular prints, and his great longevity was widely accepted for well over two hundred years.

Taylor, in his celebratory poem *The Olde, Olde, Very Olde Man*, observed in 1635:

> Some may object, that they will not believe
> His Age to be so much, for none can give
> Account thereof, Time being past so far,
> And at his Birth there was no Register.

But, on the basis of Parr's various historical recollections and anecdotes, Taylor was confident that he was as aged as he claimed. He explained that the old man had prolonged his life by eating good Shropshire butter and garlic, enjoying clean air and hard out-door labour, whilst avoiding wine, tobacco and the pox.[58]

'Old' Parr was conveyed to Westminster to meet King Charles. He was a popular sight along the route to London, with the common people crowding around to see this remarkable example of human longevity. As Taylor recorded, the curious came 'in such multitudes' to see Parr that 'the aged man [was] in danger to have bin stifeled', so keen were people 'to gaze after novelties'.[59] The illiterate may not have been able to read Taylor's valedictory poem, but no doubt many heard it, and for centuries to come Parr's cottage would be a local tourist attraction.

Arundel entertained the very old man in his London home, and displayed him to a fascinated public at the Queen's Head Tavern in the Strand. Parr, however, soon took ill. In November 1635 he died at Arundel's house, at the weathered age of 152 years and nine months.[60] On the King's command, the physician William Harvey, famous throughout Europe for his discovery of the circulation of the blood, carried out an autopsy – at that time a relatively new practice in England.[61] Harvey's examination of Parr's 'organs of generation' confirmed the sensational report that, even at the age of one hundred and twenty, the old man had had intercourse with his wife. Nor did Harvey find any great signs of ageing in the old man's other organs; having examined the stomach and intestines, the great physician deduced that by 'living frugally and roughly, and without cares, in humble circumstances', Parr had thus 'prolonged his life'. Indeed, Harvey found that 'all the internal organs seemed so sound that had he changed nothing of the routine of his former way of living, in all probability he would have delayed his death a little longer'. Harvey blamed what actually appeared to be Parr's *premature* death on the smoky atmosphere of London compared to the fresh country air of Shropshire, compounded by the old man's sudden change of diet to one more rich and varied than those plain foodstuffs to which he had been long accustomed.[62]

In recognition of his longevity, Parr was buried in Westminster Abbey. His great age was given additional gravitas by Harvey's autopsy report, which appeared in the august pages of the Royal Society of London's *Philosophical Transactions* in 1668. Unwittingly, Harvey gave Parr's longevity an official stamp that, as we shall see in Chapter 4, even sceptically minded Victorians found hard to shift. As we shall also discover, Parr's name recurs repeatedly over the next two and a half centuries, as a proven example of man's potential lifespan and of what could be achieved by living a frugal, simple, outdoor life. Harvey cannot be entirely blamed for this. There appear to have been few sceptics, and in 1661 John Evelyn – another scholarly man who had studied at Oxford and had toured the continent – happily used Parr's seemingly untimely death as clear evidence of the harmfulness of London's polluted, smoky air.[63]

The post mortem of another old Englishman by another prominent anatomist added further credibility to Parr's great age. In April 1706, an elderly, emaciated button-maker named John Bayles died in Northampton – supposedly in his 130th year. Whilst still alive, Bayles had come to the attention of a local doctor, James Keill (1673–1719). Keill was no ordinary country practitioner, however. A graduate of the University of Edinburgh, he had studied medicine at Leiden and had lectured at Oxford and Cambridge Universities before settling in Northampton. When Bayles died, Keill undertook an autopsy, sending his report to the Royal Society of London, which published it in its *Philosophical Transactions*. Keill observed that there was 'no Register so old in the Parish' where Bayles lived by which to date his birth, but the oldest locals – 'of which some are 100, others 90, and others above 80 years' – all agreed that Bayles had 'been old when they were young'. Keill recorded that, although their accounts differed 'much from one another', they concurred 'that he was at least 120 years' old. Bayles's claim that as a twelve-year old he had been at Tilbury Camp, where Queen Elizabeth had addressed her army before the Spanish Armada in

1588, was proof enough for Keill that Bayles 'must have been 130 when he dyed'.

As Keill noted, his autopsy 'agrees with that given of old *Parre* by the famous *Harvey* in most particulars'.[64] Keill attributed both men's longevity to the size and strength of their heart and lungs – though he noted that his conclusions would only be confirmed by more 'Dissections of old persons, and these are not numerous enough to ground any thing certain upon'. Nevertheless, it seemed likely to Keill that, if physicians were to find 'Rules for the preventing the ill consequences of extream old Age', the preservation of the original elasticity and softness of the bodily fibres (what we would call tissues) in their youthful state would be key.[65] The editor of the *Philosophical Transactions* may have been more sceptical than Keill: he noted that Bayles was 'reputed to have been 130 years old'.

Surprisingly, neither Parr nor Bayles were the oldest men living (or dying) in early modern England. In 1670, Henry Jenkins of Ellerton-upon-Swale, North Yorkshire, died at the remarkable age of 169. Like

Figure 1.3 Henry Jenkins: etching by Thomas Worlidge, after Walker, 1752. According to the physician Tancred Robinson, at 169 Jenkins was 'the oldest Man born upon the Ruines of the Postdiluvian World'.

Bayles, Jenkins had an historical event – as well as a number of other seemingly verifiable anecdotes – by which the authenticity of his great age could be ascertained: he claimed to have carried arrows to the battle of Flodden Field in 1513, when (he said) he was a boy of twelve. According to an account that Dr Tancred Robinson forwarded to the Royal Society, and which was again published in the *Philosophical Transactions*, four or five old men of Ellerton, 'that were reputed all of them to be an Hundred Years Old' or so, all agreed that Jenkins 'was an elderly Man ever since they knew him; for he was born in another Parish, and before any Register were in Churches'. Jenkins was, therefore, Dr Robinson noted, 'the oldest Man born upon the Ruines of the *Postdiluvian* World'.[66]

Yet, almost unbelievably, even Jenkins had a rival. In an essay published in 1683, Dr Edward Madeira Arrais of Lisbon, a physician to King John IV of Portugal, recounted reports of an Indian man who 'lived above three hundred and thirty five years', his teeth falling out and regrowing, his hair turning grey and then returning to black. Arrais told his readers that several Portuguese travellers and officials had 'at their Return from the East Indies assured me they saw him alive'.[67]

These were impressive, almost wondrous, records. They illustrate how the boundaries of old age in the early modern period were inconclusive, how the hope that life *might* – indeed *could* – be prolonged for decades well beyond seventy or eighty years was not improbable. But these boundaries were still far more limited than the near thousand-year lives of the patriarchs. By comparison, even these contemporary exemplars of longevity had aged and died prematurely.

It is crucial to realize that the quest to return human lifespans to patriarchal lengths included the long retention of *youthfulness*. Parr at 152 was still old and blind. He was remarkable, yes; but he was not even close to outliving Methuselah. It was believed that the patriarchs had grown old more *slowly* than contemporaries: Methuselah at ninety would not have looked like a seventeenth-century man looked at ninety. The patriarchs were fit and strong and healthy and youthful for far, far longer; they even had children when they were centuries old. To prolong life, ageing had to be slowed. This was the key to early modern longevity. Ageing, as Francis Bacon put it, had to be *prorogued*. But this, as he complained, was something 'that no Physitian hath handled ... according to the Merit of the subject'.[68]

The seemingly precipitate process of modern-day senescence therefore raised two pressing questions. First, what exactly explained the
rapid ageing and considerable diminution of early modern human
lifespans? And, second, how could that process be slowed or reversed
and life prolonged? As we shall see, there was a number of possible
answers. Before we can begin to understand the quest for prolonging
human flesh, however, we need to ask another basic – though
surprisingly awkward – question. What exactly *is* old age, and what
were thought to be the causes of senescence? Indeed, what is life itself,
and death for that matter? They are difficult questions, and ones that
we will return to throughout this book.

§

'Thy beauty shall no more be found,' the Yorkshire MP Andrew
Marvell wrote around this time in his taunt to his coy mistress, 'Nor,
in thy marble vault shall sound / My echoing song: then worms shall try
That long preserved virginity, / And your quaint honour turn to dust, /
And into ashes all my lust.'[69] But it did not take a poet to recognize that
the days of our youth are short. As Edward Maynwaringe wrote
rhetorically to his patients in 1670: 'I know every one of you would live
long, but especially in *health*: you would fain *continue* and *prolong* your
youth; your *beauty* and *ability* of *parts*; you are *frighted* at the thoughts
of a *wrinkled face*, or a *restless bed*; an *unwholsom diseased* body, and a
decrepid loathsom old *Age*.'[70]

The London physician John Smith, writing in 1666, described old
age as 'these evil days, and unpleasant years'. Yet he did not think it
possible to 'exactly put the terms of any mans old age, so as to say he is
now old at this present moment, but was not so before; for it is that
which creeps on by steps and degrees, as the shadow upon a Dial'.[71] No
prude or puritan, Smith was only too aware of the beauties of youth
and the horrors of growing old. His was the post-puritan era of the
Restoration court, of Charles II and his numerous mistresses, and
Smith felt free to admire the bodies of attractive young women, to see
in sexuality both the signs of youthfulness and the evidence of ageing:
'the stiffness, lively colour, and freshness of the nipples,' he recorded,
'the smoothness, fairness, elevation, and towering of the breasts, as it is
called in Scripture, *Her breasts are as towers*'. These physical

characteristics, like the 'appetite, aptness, and ability for Copulation' and the 'Inflation, and Turgescency of the Seminary vessels both *preparatory*, and *ejaculatory*' in men, were the 'excellencies of Nature' in our prime. Sex in these middle years is natural and easy. But 'as Age enfeebleth a man the grindings are weaker, and the several voices of them more submiss; wherefore it doth naturally follow, that in decrepit age, all the before mentioned indicatours of strength and perfect Concoction must be depraved, diminished, or abolished'.[72] As a woman ages, her spirits fail and her organs are 'made unfit for their Functions'. Hair turns white, joints and heart tremble, speech becomes difficult, there is 'failing of the eyes, and astonishment, paleness of the face, horrour, gnashing of the Teeth, involuntary Emission of Excrements', and death soon follows.[73]

If these were the obvious signs of ageing, harder to calculate was the exact moment of death. 'Ye way to know whether a man be dead or not', wrote John Ward, a graduate of Oxford University with a medical interest, was to 'lay a feather upon his lips; if it move hee is alive, if not, dead'.[74] Less subtly, Sir Francis Bacon recalled seeing a traitor disembowelled alive, 'whose heart being cast into the fire, leaped five foote high, and afterward lower for the space of seaven or eight minutes'. And there was another report Bacon had heard, of 'a man executed and embowelled' who, 'after his hart was pluckt out, and in the hang mans hand, was heard to utter three or four words of his prayers'.[75] The man was Sir Everard Digby, hung, drawn and quartered in 1606 for his part in the Gunpowder Plot. When his body was cut down from the gallows and eviscerated, Digby was still alive. When the executioner held up the bloody internal organs and cried out to the crowd, 'Here is the heart of a traytor!', John Aubrey recorded that it was 'credibly reported' that Digby responded, 'Thou liest!'[76] Bacon did not dismiss this remarkable account, which he found 'more likely' to be true than some other stories he had heard.[77] Clearly, the moment of death was not as immediate as might be supposed. It was not something that happened in an instant; like growing old, it was a process – albeit a swifter one.

In spite of being unable to identify the exact moment of death, Bacon defined it as the 'privation or depriving of the Sense and motion of the Heart, Arteries, Nerves, and Sinewes, inability of standing upright, stiffeness of the Nerves and limbs'. These signs of decease were followed in short course by 'coldnesse, putrefaction, and stinke'.[78]

If these were the clear indicators of the geriatric and terminal processes, what exactly was their *cause*? Even in the seventeenth century, the chief authority on this subject was the ancient Greek philosopher Aristotle. Back in the fourth century BC, he had written two short essays exploring these questions: 'On longevity and shortness of life', and 'On youth and old age, on life and death, and on breathing'. A clear sign of life in most birds and mammals is warmth. But whence does this heat originate? In Aristotle's opinion, it was the soul – which he located in the heart – that gave this warmth to living things. What the soul was exactly, Aristotle did not explain: but from the fiery soul came the very spirit of life.

The soul gives heat to the heart, and it, in turn, warms the blood, which carries that warmth around the body. If the blood ceases to be warm, Aristotle explains, 'death always ensues', for 'the soul is, as it were, set aglow with fire in this part . . . Hence, of necessity, life must be coincident with the maintenance of heat, and what we call death is its destruction.'[79] In turn, this soul–heat of the heart depended upon the fuel of nutrition, which comes from food. Without food, 'the fire fails', and all living things die. Death, Aristotle explained, could also come from *too much* heat, and this was the purpose of breathing. Air is inhaled into the lungs to cool or 'refrigerate' the flame of the soul. Thus, if we stop breathing, the body quickly overheats, and we die.[80] There was, therefore, a circular, symbiotic relationship between heat from the heart and refrigeration from the lungs. The body has to keep a healthy balance between these two states.

There was, however, a problem: gradually the body's life-giving heat consumes the lungs, and slowly dries out the flesh. (Though Aristotle did not use this metaphor himself, later Aristotelian philosophers would liken the heat-giving soul to the flame of a candle or lantern, the body being the fuel that is slowly consumed by that fire.) Aristotle thus explained that '[y]outh is the period of growth of the primary organ of refrigeration, old age of its decay, while the intervening time is the prime of life'. Death in old age 'is the exhaustion due to inability on the part of the organs, owing to old age, to produce refrigeration'. Gradually, like the candle-flame consuming both the wax and the wick, the spirit of life consumes its fuel and fades. For Aristotle, death in old age is thus both 'natural' and 'painless': 'It is just as though the heart contained a tiny feeble flame which the slightest movement puts out.'[81] Ironically, the source of life – heat – is also the very cause of death.

To the slow, inexorable and seemingly inevitable process towards old age and death, various external factors could be added which would affect the speed of its progress. These all either hastened ageing by accelerating the drying-out process, or alternatively could slow it down. These factors were important, and we will return to them repeatedly in the course of this book: they included such things as excesses or errors (or potential benefits) in what we eat and drink; the ill or beneficial effects of air and temperature (it was thought that people aged faster in cold climates, but lived longer in hot ones); and physical activities, which were clearly significant: work and exercise were considered to be good, whilst sex was bad – particularly for men. As Aristotle explained, the expulsion of a man's 'seed' hastened the ageing process, because its loss 'produces dryness'. This was why 'females live longer than males if the males are salacious'. But 'as a general rule males live longer than females, and the reason is that the male is an animal with more warmth than the female'.[82]

Classical physicians established several principles for leading a long, healthy life. These focused on the so-called 'six non-naturals' laid out in the Hippocratic corpus between the fourth and the fifth centuries BC, and standardized by Galen in the second century AD. Control these and the body would remain healthy and free from sickness and disease. The non-naturals were: air; diet; exercise; sleep; passions of the mind; and bodily excretions (which included sweat, urine, faeces, vomit, saliva, phlegm, menses and semen). Keeping healthy, according to these rules, meant particular attention should be given to diet. As one seventeenth-century physician memorably put it, through insufficient consideration of what they ate and drank, many people 'dig their Graves with their Teeth', and thereby 'cut off the Thred of their Lives sooner than is required by God, or Nature'.[83]

Another contemporary, Dr James Hart, noted that in antiquity there were some people 'who by meanes of diet' promised 'the perpetuity of mans life, and of a mortall man, to make him immortall'. Hart pointed out the criticism made by the great Greek physician of the Roman world, Galen, in the second century AD, of a philosopher 'who promised immortality to all such, whose education he had from their tender yeeres undertaken.'[84] Hart, like Galen, dismissed such an idea:

Galen broadly considered old age a natural phenomenon, and not an 'illness' that could somehow be 'cured'.[85] He suggested that, since it was 'impossible for the body to escape the natural road to dryness', it was 'inevitable for us to grow old and perish'.[86]

Though untreatable, old age was not thought of as coming any sooner in the classical world than in Bacon's day; nor was it necessarily considered a burden. Various authors in the Hippocratic corpus saw the 'green springtime of decline' as not starting till seventy, with old age only beginning at seventy-five.[87] In his famous essay *De senectute*, written in his sixty-third year and shortly before his murder in 43 BC, the Roman statesman Cicero idealized old age as the pinnacle of life. At least two Roman emperors – Augustus (63 BC–AD 14) and Tiberius (42 BC–AD 37) – lived into their late seventies, as Francis Bacon pointed out in his catalogue of the aged.

Over the centuries, classical and Islamic physicians such as Galen and Avicenna amended and augmented Aristotle's theory of the role of heat and moisture in life and ageing, and debate continued as to whether the latter was to be considered an illness – curable or otherwise. The medieval viewpoint on life and death was based on this long medical debate, embellished by Christian theology and eschatology. In the thirteenth century, the influential (and impressively long-lived) English philosopher Roger Bacon (*c.*1214–1292) wrote a treatise on longevity; an edition was published in 1683 as *The Cure of Old Age, and Preservation of Youth*. There he explained that, as humans grow older, they gradually lose both their 'natural heat' – their inner vitality, and their 'natural moisture' – that is, the softness and suppleness of youth, evidenced in the smooth flesh of a child but so clearly absent in the dry, wrinkled skin of the geriatric body. This 'natural heat', Roger Bacon explained to his medieval readers, began to diminish 'after the time of Manhood, that is, after forty or at most fifty Years'. As a man grows older the cycle of ageing progresses ever faster, so that he would be 'sooner turned from Old Age to decrepit Age, than from Age to Old Age'.[88]

§

If these were the causes of old age, how could the detrimental effects be slowed, and life extended? A problem with the Aristotelian theory was that it did not fully explain how the patriarchs had lived almost a thousand years. Of course, Aristotle was a pagan Greek who quite probably knew nothing about Hebrew myths of origin, so there was no reason at all why he should have provided an answer to this puzzle. To inquisitive early modern Christians, however, the reason why we age so quickly and our lives are so foreshortened was a conundrum requiring an answer.

For the ancient Jewish scholar Josephus, it was the fact of their virtuous behaviour, together with their proper diet, that accounted for the patriarchs' longevity.[89] According to Roger Bacon, however, there were at least three explanations as to why humans aged more quickly than in previous epochs. First, there was the fact that, as 'the World waxeth old, Men grow old with it: not by reason of the Age of the World, but because of the great Increase of living Creatures, which *infect* the very Air'.[90] According to Roger Bacon's medieval interpretation, there was little that could be done to recover the long lives of our ancestors. As the vitality of the earth waned, so too did the things living upon it.[91] As William Russell, chemist-in-ordinary to Charles II, wrote in 1684 echoing Bacon: 'it appears that the World it self waxeth old, the Powers thereof are much altered, [and] all the external Virtues of its superficies are declined'.[92]

The earth, it therefore seemed to some, was drawing inevitably, inexorably, to a close. This weary planet, like an aged mother, could not so easily sustain the creatures living upon it; time was coming to an end, human flesh crumbling. The troubled political events of the seventeenth century – together with the apparent increasing incidence of diseases such as tuberculosis, smallpox, scurvy and rickets – seemed to suggest as much. As Richard Browne observed in 1683 in his notes to Roger Bacon's book on prolongevity, 'we must conclude the World is in its testy old Age', and the Second Coming of Christ was nigh.[93]

Even if it was not accepted that the world was ageing and would itself die, it could be argued that it was not so fertile as it had been in the earliest times, when men had lived healthy lives of hundreds of years upon it. An obvious explanation for this reduced fecundity was the Flood – an argument put forward by, among others, Sir William Vaughan (*c*.1575–1641) in his remarkably popular *Directions for Health, Naturall and Artificiall*. Vaughan explained that the 'principall reason' men had lived longer before the Flood was that the world had then been in much better state: 'the earth in those dayes was of greater efficacie to bring forth necessaries for mans use, than it is in this crooked and out-worne age. The soyle was then gay, trim, and fresh: whereas now by reason of the inundation . . . it is barren, saltish and unsavorie.'[94] This was an argument Sir Francis Bacon supported. He blamed the cumulative effect of this Great Deluge, together with other smaller floods, long droughts and earthquakes, for the steady decline in human lifespans since Noah's day. Somehow, he believed, these events had made the land less fertile, or the air less pure.[95] A more physical explanation was suggested by the ingenious Robert Hooke: he proposed to the Royal Society in 1668 that friction had had the effect of slowing the earth's rotation, thereby increasing the length of the days and years: a patriarchal year had thus been much shorter than a modern one.[96]

According to Dr Edward Madeira Arrais, Adam and Eve had lived so long because of some property to be found in the fruits of the trees in Eden. He assessed differing opinions as to whether the Tree of Life had actually conveyed immortality or only a long duration of life, and whether it was necessary to eat only once of the tree to obtain its benefits or one had to eat of it repeatedly.[97] Arrais decided that 'the very long Life' of Adam and the patriarchs should be ascribed to the 'Qualities' of this and 'other wholesome Trees growing in Paradise, which were either in Fruits to be eaten, or transfused to the ambient Air . . . and then they were communicated by Food and Air to the Bodies of our first Parents, and from them again in Seed and Blood to their Children'. It was by this reasoning that some ascribed the longevity of Arrais's ancient Indian to his eating the fruits he found floating in the Ganges. Arrais did not believe these fruits had come from the Tree of Life itself, but he thought they were certainly 'very wholesom Fruits of some other Trees', which kept off diseases.[98]

Roger Bacon suggested two further explanations for man's shortening lifespan: 'our *Negligence* in ordering our Lives, and That great *Ignorance* of the Properties which are in things conducing to Health, which might help a disordered way of Living'. The latter had made little progress by the seventeenth century. The knowledge of medicines was still uncertain and potentially dangerous, but we will look much more closely at this in the next chapter. Negligence in how we lived our lives was far more pressing, and something about which lots of things could be done – as had long been realized. It was here that the greatest attention would focus.

As we have seen, Thomas Parr was an excellent example of what could be achieved by eating simply and labouring hard in clean, country air. There was also a famous gentlemanly example: in 1558, a ninety-one-year-old Venetian nobleman, Luigi Cornaro, published an influential book on the role diet could play in prolonging life. Cornaro (whose great age is accepted as accurate by modern historians) lived long enough to update his *Discorsi della vita sobria* numerous times before his death, at the age of ninety-nine, in 1566.[99] As a young man, Cornaro had indulged in rich food and luxuries, but at forty he realized that if he continued in such a fashion he would live only little longer. So he abandoned his diet of strong wines, raw fruits and baked meats dressed with rich sauces. Instead, he adopted a frugal daily regime consisting of only twelve ounces of bread, soup, egg yolks and meat, and some fourteen ounces of wine. There were to be no more late nights, over-strenuous exercise, women, or grievous, melancholy thoughts in his life. A man who took such care with his health, Cornaro observed, 'prolongs his Life to above a Hundred Years, spares him[self] the Pain of a violent Death, sends him quietly out of the World'. But, he lamented, 'most Men suffer themselves to be seduced by the Charms of a Voluptuous Life' – and they reaped the consequences with foreshortened lives. He reckoned that gluttony and excess robbed Italy annually 'of more Inhabitants, than Pestilence, War and Famine'.[100]

Cornaro warned, however, that 'it must not be suppos'd' that his rules would 'make a Man immortal'. This was impossible, for all things come to an end. Nonetheless, all men ought to close their days 'by a natural Death, that is, without any Pain; as they will see me dye, when the radical Moisture shall be quite exhausted'.[101] Sir William Vaughan likewise opined that '[t]o live forever, and to become immortall here on earth, is a thing impossible'. But he, too, observed that there were certain things that could be done 'to prolong a mans life free from violent sicknesses'. Like many other commentators, he looked back to the example of the patriarchs, observing that they had been more continent in their diets: 'they knew not our dainty eates, our march-panes, nor our superfluous slibber sauces; they were no quaffers of Wine or Ale, nor were they troubled with so many cares, nor with passions of Envy and Malice'.[102] It was these bad habits that had undermined the human frame.

Alcohol was often blamed for shortening human lifespans. The Bible makes it clear that, following the Flood, Noah was the first person ever to cultivate grapes and make wine. It did not take a genius to note that it was also since the age of Noah that lifespans had begun their sudden decline. So, if it wasn't the fault of the Flood, maybe it was the fault of alcohol – or both?

Not everyone held this view. Roger Bacon had recommended drinking red wine for its health-bringing properties, whilst in 1638 the London physician Tobias Whitaker devoted a whole book to the subject. In his *The Tree of Humane Life, or, The Bloud of the Grape*, he set out to prove 'the possibilitie of maintaining Life from infancy to extreame old age without any sicknesse by the use of Wine'. According to Whitaker, both Parr and the Countess of Desmond were 'extraordinary examples' of what could be achieved by following 'that puritie of Principles' which had helped the patriarchs to 'have exceeded the age of nine hundred yeares'.[103] Though he did consider death inevitable, 'because the heart cannot bee made moister', Whitaker did not think it 'wise . . . in a negligent way to betray our lives to death before the time', and he accepted 'the possibility of extending [life] to extreame old age'.[104] Not surprisingly, Whitaker's argument was well received in some quarters: his book was republished several times and translated into Latin, with editions appearing in Frankfurt and The Hague. In 1660 he would be appointed a physician-in-ordinary to the Restoration household of Charles II. Only four years later, however, Whitaker was dead. His method for longevity had not paid off: he was probably only in his mid-sixties.[105]

If attitudes to wine were ambivalent, meat was a prime culprit, singled out in the bid to understand why modern humans aged fast and died young. Again, the Bible made it clear it was only *after* the Flood that we became omnivores. God had told Adam in Eden: 'Behold, I have given you every herb bearing seed, which is upon the face of all the earth, and every tree . . . to you it shall be for meat.'[106] It was only after the water subsided from earth's face that God told Noah: 'Every moving thing that liveth shall be meat for you; even as the green herb have I given you all things.'[107] The pious polymath Tomas Tryon (1634–1703) was an early advocate of vegetarianism, advising his readers to 'avoid all excesses in foods and drinks, either in quality or quantity, to eschew things derived from violence, and therefore be considerate in eating Flesh and Fish, or any thing not procured but by the death of some of our fellow creatures'[108]. In his *A Way to Health, Long Life and Happiness* (1691) he pointed out that the patriarchs had lived long lives, free from sickness, partly on this account. His followers included John Evelyn, who, in a discourse on salads published in 1699 and written when he was nearly eighty, pointed out that the patriarchs had lived prodigiously long lives on the 'wholesomeness of the Herby-Diet'.[109]

The various faults in diet and behaviour were seen as hereditary errors, passing from parent to child. Again, it did not take too close a reading of the Bible to notice that, although God had told Adam and Eve *before* the Fall '[b]e fruitfull and multiply, and replenish the earth, and subdue it', it was not till *afterwards* that 'Adam knew Eve his wife: and she conceived, and bare Cain'.[110] Sir John Pettus thus observed in 1674 that Adam first become aware of sex only by eating the fruit of the Tree of Knowledge of Good and Evil. (This was the reason why writers such as Simon Foreman suggested Adam and Eve had not had genitals before the Fall.) Pettus wrote that it was only with the discovery of sex that 'Man hath the reward of his Libidinous disobedience, his body being so full of Disease and Infirmities, that the means of propagation seems to beget more Diseases than Children'.[111]

Not everyone agreed. Tobias Whitaker suggested that, if 'rightly used', sex (like wine) was a good thing: it could 'exhilerate the minde, cheare the spirits, refrigerate the body, and cause sleepe'.[112] But to those of a more puritanical disposition, little good could come of venery: it exhausted the store of radical moisture, with sexually transmitted diseases an added, ever present (and potentially fatal) danger. That sex is bad and shortens life is an argument that resounds down the centuries.

As we have seen, prior to the Fall Adam and Eve were teetotal, celibate vegetarians. The subsequent, post-lapsarian ill effects of all that meat eating, wine drinking and fornication were passed on in a degenerative cycle which, five and a half thousand years later, had left us pale imitations of our first parents. In these circumstances, to prolong our lives would involve some considerable reversals in the ways we lived.

The unknown author of *The Way to Bliss*, whose manuscript was published in 1658 by the antiquary Elias Ashmole, proposed a novel experiment that distils many of these ideas. He considered it unreasonable that the human being, the most perfect of God's creations, should live a shorter life than other animals, such as the elephant (which, Aristotle claimed, lived for three hundred years). He therefore suggested that 'if a company of pickt and lusty *Men* and *Women* would agree to live together in some wilde, open, clear and sweet *Air*, scatteredly like a Country Village, and not like a close and smothered City, (which one thing prevents a thousand Diseases and Deaths alone)'; if they agreed to have sex only in order to procreate, 'and not for Pleasures sake'; if they agreed 'to bring up their Children in Labour and Hardship, mingled with much Mirth and Sleep together'; and if they would take no medicines other than when in the greatest danger, and would never drink wine or eat cooked foods:

> If these things, I say, were duly kept and performed, I am fully perswaded within three or four Generations and Off-springs, it would come to pass, that we should see this People prove a Nation of Giants, not onely passing the age of Beasts, and the bounds of *Long Life* afore-said, but wholly recovering and restoring all the Blessings of the first estate of *Body*.[113]

It was a novel idea – though unfortunately there is no evidence that this utopian experiment was ever attempted.

§

Having explored the wider context of ageing in the early modern period, we can look more closely at the schema devised and advanced by Sir Francis Bacon in his *History of Life and Death*. The ideas on ageing and death proposed by classical physicians and philosophers had changed little by his day.[114] Although Bacon claimed to reject the theory of radical heat and moisture, it in fact formed the basis of his own theory. The chief difference was that it was now more widely held that the human soul – the source of the body's heat – resided in the brain, and not the heart. So, although Bacon held that there were 'many *Paths,* which lead to *Death*',[115] the principal route to 'natural' death was the drying out of matter caused by the heating force of those same 'vital *Spirits*' that give the power of life and movement. These spirits, located in the brain and akin to 'the substance of *Flame*', were 'the Master-workmen of all Effects in the *Bodie*': as well as giving life, they instigated all mental and physical operations. But, along with these vital spirits that existed in living things, there were also lifeless spirits that were present in all matter, animate and inanimate. They, too, contributed to decay; and it was important to Bacon to note that 'to procure long Life, the Body of Man must be considered; First, as *Inanimate,* and not *Repaired* by *Nourishment:* Secondly, as *Animate,* and *Repaired* by *Nourishment*'.[116]

Thus, although Bacon dismissed Aristotelian and scholastic theories of the soul, he still identified it as an airy, fire-like spirit.[117] (For the seventeenth-century physician and controversial vicar John Pordage, these spirits had a theological interpretation: they were 'the fiery deity of Christ', who 'did mingle and mixe it selfe with our flesh, and was in the center of our Soules, burning and consuming'.)[118] But Bacon placed greater emphasis than his predecessors on the role of these 'spirits', as well as on the effects of digestion.[119] If the wasting heat of the spirits could be checked, life would be prolonged; he also believed, importantly, that damage could be *repaired* – that new 'fuel' (or 'nourishment') could be added to the fire of life.

In this respect, Bacon's principal matter of disagreement with Aristotle's theory was the suggestion that ageing and death were *irreversible.* Instead, and most importantly, he claimed: 'That which may bee repaired by Degrees, without a Totall waste of the first Stocke, is potentially eternall.' This, he pointed out, was the case with the flame that had burnt in the Vestal temple in Rome: through constant replenishment, it was never allowed to go out.[120] Perhaps the same was possible by tempering and refuelling the flame of the human spirit?

Bacon's second crucial point was that the ageing process is *unequal*: certain parts of the body, such as the spirits, blood, flesh and fat, 'are, even after the Decline of yeares, easily repaired'. It is the 'drier' parts, such as the sinews, veins, arteries, bones, cartilages and bowels, which are 'hardly Reparable'. As these latter parts decay, and in the end 'utterly fail', they ultimately take the whole fabric of the body with them. Solve these problems, Bacon reckoned, and you conquered death.[121]

His rules for prolonging life were thus simple and two-fold:

1   Inhibit the ageing process by reducing the heating effects of the spirits.
2   Repair local damage before it spreads and 'a generall ruine follows'.[122]

Laid out in its first few pages, this was the principal thesis of Bacon's *History of Life and Death*. However, though his basic rules might seem simple, he explained that 'so great a Worke as the Stopping, and Turning back, the powerfull Course of Nature' was not to be achieved without great effort. Certainly, simply taking some morning medicine or 'precious Drug' could not alone do it. He assured his readers

> that it must needs be, that this is a work of labour; And consisteth of many Remedies, and a fit Connexion of them amongst themselves; For no Man can bee so stupid, as to imagine, that what was never yet done, can bee done, but by such wayes, as were never yet attempted.[123]

As the stomach was 'the *Master* of the *House*', and as it was from food that the juices of the body were formed, diet was crucially important to Bacon's method for prolonging life.[124] He considered Luigi Cornaro's meagre intake too frugal to be worth pursuing, and he wasn't going to starve himself in the attempt to live long. He encouraged consumption of sweet things (such as fruits and honey), fatty, roasted meats, and salads with marigold leaves and betony flowers; wines and ale supplemented with various beneficial additives were also to be drunk, varying according to time of year.[125]

Diet was efficacious only as a part of Bacon's broader regimen, however. Meals were to be followed by rest, with sound sleep encouraged. Regular, gentle exercise was important: 'leaping', riding, shooting and bowls he recommended, though tennis, running and dancing were 'too Nimble' and overexcited the spirits.[126] Control of the '*Passions* of the *Minde*' was equally important, though he felt that amongst physicians there was 'a deepe silence' on this subject: joy, hope, contemplation and reflection all encouraged long life, whilst fear, anger, sadness and envy debilitated the spirits.[127] Good, clean air and morning walks in country gardens, smelling the scent of flowers, were particularly helpful 'for the Comforting of the *Heart*'.[128]

But these were just the basics. His first aim, to temper the damaging 'eager Flames' of the spirits that aged the body, would be achieved by the 'cooling' effects of drugs such as opium and tobacco. The 'knot' or paradox here, Bacon noted, was that the spirits, by escaping the body, age it; but by being kept *in* the body, they heat it, and cause decay. The aim of his method was to keep the spirits in, whilst at the same time keeping them cool and diffuse.[129] To this end, Bacon recommended, from youth onwards, an annual 'Opiate Diet'. By opiates Bacon meant both opium itself, and other 'hot' drugs such as henbane, mandrake, hemlock, tobacco and nightshade. A light concoction 'of a superior kind' was to be drunk every other morning for a fortnight in spring. 'Opiates' of a weaker nature, such as saffron, ambergris (a secretion from the intestines of sperm whales), nutmeg and rose water, could all be taken more regularly, '[a]nd they will be very effectuall to prolongation of life'.[130] The animal spirits could also be 'refrigerated' with 'nitre' (saltpeter, or potassium nitrate, a naturally occurring mineral used in medicine as a diuretic, as a preservative of food, as a fertilizer and in the manufacture of gunpowder). This could be mixed into food or beer, and 'is of prime Force, to long Life'.[131]

The best way to exclude the 'predatory' effect of air was by living in dry caves and on mountains. This inhibited the escape of the spirits, and the inhabitants of Barbary, Bacon noted, lived up to 150 years on this account.[132] But Bacon, a man of business, was hardly going to move to north Africa and live in a cave. As an alternative, one might dress the skin with olive or almond oil, thus more conveniently (and aromatically) closing the pores through which the spirits escaped. This, he supposed, would work in much the same way paint and varnish protect wood from decay, and was analogous to the way that insects caught in amber 'doe never after corrupt, or rot, although they be soft and tender Bodies'. Though he admitted that this 'anointing' with oils had its 'inconveniencies', he considered it 'one of the most potent Operations to long Life'.[133]

Astringent washes could also be used to make the skin harder; hot baths, however, were to be avoided, as they opened the pores.[134] To cool the blood, he recommended enemas and lukewarm baths of fresh water and oils; and the blood itself could be firmed up against the actions of the spirits by occasionally taking powdered gold, pearl, coral or precious stones in white wine, or alternatively decoctions of oak, vine or rosemary.[135] He added that, if it did not 'seeme to us Sluttish and Odious', bathing in human or animal blood would help keep the flesh supple. Those opposed to bathing in blood could try milk, egg yolks, wine, saffron or myrrh instead. Ideally, such a bath would be preceded by a massage and followed by the application of a thick oil of mastic, myrrh and saffron, to help the bath's 'Moistning Heat' penetrate the flesh. This oil should be left on for twenty-four hours, the process being repeated every fifth day for a month. Bacon noted that he had not actually tried this 'Experiment' himself, but offered it as a light for others to follow.[136]

§

These methods, some tried and tested, others merely hypothetical, were Bacon's principal means of preventing ageing and encouraging restoration. But did he practise any of them himself?

In his essay 'Of Death', Bacon wrote: 'It is as natural to die as to be born', and he did not expect to live prodigiously long.[137] In the second part of his *Novum Organum* (1620), he acknowledged that 'we entertain no hope of our life being prolonged to the completion of the sixth part of the Instauration'.[138] But he took great care of his health. His letters from his estate at Gorhambury reveal that 'by means of the sweet air of the country' he obtained 'some degree of health' in his last years.[139]

William Rawley, Bacon's chaplain and secretary from 1618 to 1626, recorded that Bacon observed many of his own rules for prolonging life: these included 'rather a plentifull, and liberall, Diet, as his Stomack would bear it, than a Restrained [one], Which he also commended, in his Book, of the History, of Life, and Death'. In his youth, Rawley noted, Bacon had favoured lighter meats, such as fowl; 'But afterward, when he grew more Judicious; He preferred the stronger Meats', which 'bred the more firm, and substantiall Juyces, of the Body'. Rawley also records that Bacon 'took each morning in his broth about three grains of Nitre, for thirty years together, next before his Death. And for Physick, he did, indeed, live Physically, but not miserably.' About once a week he drank 'a Macreation of Rhubarb' – a purgative – infused in wine or beer, before dinner or supper, 'that it might dry the Body, lesse: which, (as he said,) did carry away frequently, the Grosser Humours, of the Body; And not diminish, or carry away, any of the Spirits; As Sweating doth. . . . As for other Physick, in an ordinary way, (whatsoever hath been vulgarly spoken;) he took not.'[140] Bacon's interest in medicines was such that in 1679 a number of his supposed recipes were posthumously published among his collected works, under the title 'Baconiana medica'. They included 'A Medical Paper of the Lord Bacon's, to which he gave the Title of *Grains of Youth*'. This in turn included a medicine made from nitre, ambergris and poppy seeds, a 'preserving oyntment' made from 'Deers-suet', saffron and myrrh, and even a 'Methusalem Water' made from crayfish boiled in claret that acted 'against the Driness of Age'.[141]

Despite these remedies and preservatives, Bacon's life was not especially long, though his death, when it came, was unexpected. It certainly seems possible that in stuffing the chicken with snow Bacon was testing the twenty-eighth of his thirty-two provisional 'Moveable Canons, of the Duration of Life, and Forme of Death' – that is, '*Refrigeration,* or *Cooling* of the *Body,* which passeth some other wayes, then by the *Stomach,* is usefull for Long Life'.[142] Excepting, of course, when it kills you. His domestic attendant, William Atkins, was stunned, likening his master's 'unexpected' demise to a besieged army caught off guard: Death thus 'smites this man much skilled in warding off a blow'.[143]

§

Why exactly did Bacon consider the prolongation of human life to be so important? Although he called his times 'this autumn of the world', and though he, too, appears to have held millenarian beliefs, he rejected pessimistic views of natural history. If this *was* the earth's dotage, for Bacon it was to be a mature old age of profound wisdom and great learning, in which European scholars would pluck the final fruits of God's benevolent creation.[144] Natural philosophers would take full advantage of all that had gone before them, restoring mankind to the greatness that had once been Adam's. Only then, when this last great age of progress had been fulfilled, would the world be fit for Doomsday.

Hence Bacon confidently cited Daniel's Old Testament prophecy touching the end of days: 'Many shall go to and fro, and knowledge shall be increased.'[145] This was the 'special prophecy' Bacon believed God had directed to his own age. Such apparently unprecedented events as the discovery of the New World and the circulation of the blood, along with recent inventions such as the printing press, the telescope and the microscope, suggested to the optimist that this prophecy was now being fulfilled: all that was required to round things off was the perfecting of science. And this would include the vast prolongation of life. For perhaps Bacon also had in mind the Old Testament lines in which God promises the Babylonian exiles that, in the new Jerusalem, 'Never again will there be in it an infant who lives but a few days, or an old man who does not live out his years; he who dies at a hundred will be thought a mere youth; he who fails to reach a hundred will be accursed.'[146]

As these lines suggest, the desire to prolong life was not irreligious. And, as Bacon pointed out, St John (the '*Beloved Disciple* of our *Lord*') had outlived all the others, whilst many hermits, monks and Church Fathers had also died aged. This surely proved that longevity enjoyed God's favour. Furthermore, he explained, longer life would give us more time to do good, Christian works.[147]

For Sir Francis Bacon, therefore, the quest for physical longevity was both a pious ambition *and* an inherent feature of the 'Great Instauration' – Bacon's grand plan to overthrow traditional learning based on Aristotle and the ancients and to re-establish scholarship on new, experimental, empirical and essentially modern foundations. Not only would prolonging life be a *part* of this programme, the programme would itself be facilitated *by* the prolongation of life: the two things were synonymous.

To clarify this point, we can look at a letter written in 1651 by the English philosopher Anne Conway to her father-in-law, Lord Conway. Anne explained that human learning before the Flood had been 'very great' because Adam 'certainly was an excilent naturall philosopher as appears by his giving names to every beast and bird according to their natures'. Furthermore – and here we get a sense of what Bacon no doubt hoped to achieve – Conway added that Adam 'was contemporary' with his descendants 'for above 900 yeeres', and hence had been able to share his great wisdom. These antediluvians were further advantaged in the advancement of their learning, Conway explained, because 'the extraordinary length of their lives gave them leave to make infinit experiments, and experience is the mother of all knowledge'. As 'experiments', Conway gave the example of astronomy: whilst the antediluvian philosophers had been able to 'observe the revolution of a sphære whose circuit will not be finished in 300 yeer . . . our Life will permit us to observe very few circuits of the celestiall bodies'.

Anne Conway had no doubt as to the authenticity of these long lives. In an off-hand remark, she noted that Noah was 600 when the Flood came – 'time enough to gaine a vast knowledge'.[148] Over the subsequent centuries, as lifespans had fallen, and particularly after the destruction of the Tower of Babel, when different languages had divided men, knowledge had declined and became corrupted. It was this decay of learning that Bacon hoped to overcome, returning us once more to an adamic level of complete understanding, in anticipation of Judgement Day.

For Bacon, long life was not simply a projected *end* in the restitution of all wisdom – it would also prove to be one of its *means*. A century after Bacon died, the French writer the Marquis D'Argenson would define youth as that time 'when a man thinks himself immortal' and 'imagines projects of long duration'. D'Argenson believed that this age of optimism ended at around thirty-five.[149] But why should men age at all? If youthfulness could be maintained, would not that same confidence in pursuing 'projects of long duration' be retained, and even greater ends in philosophy and science achieved?

What Bacon appears to have been arguing for in the prolongation of life, therefore, is this. Given what a single gifted human being can achieve in one seemingly foreshortened and increasingly debilitated lifespan of seventy or eighty years, imagine what they could achieve in an almost perpetually youthful life of a *thousand* years? Thus in explaining his interest in longevity in the *History of Life and Death*, Bacon naturally cited Hippocrates' ancient dictum '*Life* is short, and *Art* long'. He then explained: 'Therefore our labours intending to perfect Arts, should by the assistance of the Author of Truth and Life, consider by what meanes the Life of Man may be prolonged.' Put simply, longer life 'affords longer opportunity of doing good Workes'.[150] He had emphasized as much in the preface to *The Advancement of Learning*: 'this Our Instauration is a matter infinite,' he had declared, 'and beyond the power and compasse of Mortality'. Moreover, the world itself was so vast that a modern human lifetime was not long enough to comprehend it all. As Robert Boyle reflected in 1663 (with an air both of awe and disappointment), 'the Laws and Works of Nature' are 'so various and numberlesse' that even 'if a Man had the Age of *Methusalah*' to spend in research, he would still run out of time before he ran out of fit subjects for study.[151]

In spite of his desire to prolong his own life, Bacon acknowledged that he was 'not unmindfull of Mortality', admitting that his '*Designe*' was not to be 'accomplisht within the Revolution of an Age only'. Rather, he delivered it 'to Posterity to Perfect'.[152] Although Bacon practised many of his own precepts for prolonging life, substantial longevity was not something he expected to be achieved quickly, or through the resources of a single individual (or in a single, 'normal' lifespan). As he explained to a correspondent, since 'these things' required 'ages for their accomplishment', they were 'plainly a work for a King or Pope, or some college or order: and cannot be done as it should be by a private man's industry'.[153] The *History of Life and Death* was not intended as a definitive work. Like the lawyer he was, Bacon was laying out the evidence for the case, heading up certain 'experiments' to be undertaken, and others to be followed up. Final judgement on the case he left pending. Bacon was well aware that he was sowing seeds for future study, rather than establishing the firm root of a new science.

One of Bacon's major (though at first unfulfilled) ambitions for perfecting his project was the establishment of societies, with sufficient funds to build permanent headquarters. These societies would perpetuate and improve knowledge and learning down the ages via the collaborative input of their members. Hence, in his posthumously published utopian tract, *The New Atlantis*, Bacon wrote of Salomon's House, an institution devoted to the collective, long-term advancement of learning. The great enterprises pursued at Salomon's House included 'the prolongation of life, the restitution of youth in some degree, [and] the retardation of age'.[154] The book was very popular, going through eight editions between 1626 and 1658 alone.

The notion implicit in Salomon's House is that, if the death of the *individual* cannot be staved off in the short term, then 'immortality' could be assured through the collective identity of the many pursuing the same philosophical and scientific goal. Bacon defined this objective in *The Advancement of Learning*, where he laid down his critical programme for a 'Seate of the Muses'. There he had advised King James that it was those who were 'fruitfull' in their offspring who had 'a fore-sight of their own immortality in their Descendants'. In this sense, he explained, the childless Queen Elizabeth had merely been 'a sojourner in the world, in respect of her unmarried life, rather than an inhabitant'. She had enriched her own times, but it was through children that one's life's work was extended, 'which succeeding Ages may cherish, and Eternity it selfe behold: Amongst which, if my affection to Learning doe not transport me, there is none more worthy, or more noble, than the endowment of the world with sound and fruitfull Advancements of Learning'.[155]

Though married, Bacon was a homosexual; he never had children and did not expect to achieve 'immortality' though physical descendents.[156] Rather, personal immortality would come to him through his work. In the decades after Bacon's death, Salomon's House became the model for numerous scientific societies: the Invisible College, the Oxford Philosophical Club and, in 1660, the Royal Society of London were all at heart Baconian institutions. In many ways, such societies encapsulated Bacon's dream of immortality.

So, even though the man might die, his work would be continued by younger successors, and his learning would survive. Books and libraries served as the bridge linking past research with future discovery.[157] Whilst Bacon may not have had children, his life was not without offspring: books, as he explained in *The Advancement of Learning*, are effectively immortal, for it is here that 'the images of mens wits, remain unmaimed . . . for ever, exempt from the injuries of time, because capable of perpetuall renovation'. Books, furthermore, are eternally fertile: they 'generate still and cast their seeds in the mindes of men; raising and procreating infinite Actions and Opinions in succeeding ages'.[158] Books are 'as ships, passing through the vast sea of time', linking 'the remotest ages of Wits and Inventions in mutuall Trafique and Correspondency'.[159]

Through books, wisdom is transported down the ages; it is books that give young men the insight and knowledge only won by long life. As the great bibliophile John Selden observed in 1618, by reading, '[t]he many ages of Former Experience and Observation . . . may so accumulat yeers to us as if we had livd even from the beginning of Time'.[160] And, as Thomas Fuller declared in 1639, studying history 'maketh a young man to be old, without either wrinkles or grey hairs; priviledging him with the experience of age, without either the infirmities or inconveniencies thereof'.[161] Elias Ashmole, writing in 1652, averred that, in gathering together and publishing his collection of alchemical works, he had acted 'as if having the *Elixir* [of Life] it selfe', and had 'made *Old Age* become *Young* and *Lively*, by restoring each of the *Ancient Writers*, not only to the *Spring* of their severall *Beauties*, but the *Summer* of their *Strength* and *Perfection*'.[162]

But institutional and literary longevity were only two facets of the Baconian project. As we have seen, physical longevity remained a clear and very real goal.

§

Did many people follow the advice laid down in the *History of Life and Death*? Though the book had sufficient readership to prompt two English translations in 1638, this is a difficult question to answer. One possible follower of some of Bacon's directives was Thomas Hobbes, one of the longest-lived seventeenth-century Englishmen whose dates have been verified: born prematurely in 1588, Hobbes lived on to an impressive ninety-one years.[163] Notably, in the 1620s he was for a time one of Francis Bacon's favoured secretaries and translators, and it is known that he held Bacon's writings in high regard.[164] He was also concerned with conserving his own health. In 1649, whilst in exile in France, he told Pierre Gassendi: 'I am in fairly good health for my age, and I am certainly looking after myself, preserving myself for my return to England, should it happen by any chance.'[165]

According to Hobbes's young friend John Aubrey, the famous philosopher followed a number of practices that he believed would prolong his life. In his youth, as Aubrey records, Hobbes was temperate 'both as to wine and women', but for the last three decades of his life he suffered increasingly from a disease contemporaries called 'shaking palsy' – which may have been what we know today as Parkinson's disease.[166] Aubrey noted that for Hobbes's last thirty years or so 'his dyet, etc., was very moderate and regular' and that in his final years he 'wase scarce able to write his name'. Furthermore, '[b]esides his dayly walking, he did twice or thrice a year play at tennis (at about 75 he did it); then went to bed there was well rubbed. This he did believe would make him live two or three yeares the longer.'[167] (Even at seventy-eight, another friend noted, Hobbes walked all morning 'for his Health', and played 'so long at Tennis once a Week till he [was] quite tired'.)[168] Every night, as Aubrey wrote, when the doors were shut and he was alone, Hobbes 'sang aloud (not that he had a very good voice) but for his health's sake: he did beleeve it did his lungs good, and conduced much to prolong his life'.[169]

It was in these last decades of his long life that Hobbes became notorious for his supposed atheism. He certainly denied that the soul was immortal, but did believe that God would revive everyone from their death sleep at Judgement Day: 'God, that could give life to a piece of clay,' he stated, 'hath the same power to give life again to a dead man, and renew his inanimate, and rotten, carcass into a glorious, spiritual, and immortal body.' His mother, Hobbes once explained, had given birth to twins, 'to myself and to fear'. As Jacques Choron has written, 'it is quite probable that the fear of death must have been not the least among his anxieties. If authentic, his last words, "I take a fearful leap into the dark", seem to confirm this view.'[170]

Hobbes's apparent attempts to escape death and elude his final reckoning with God were mocked by contemporaries. An 'Elegie' published shortly after the philosopher's eventual demise in 1679 asked:

> Is he then dead at last, whom vain report
> So often had feign'd Mortal in meer sport?
> Whom we on Earth so long alive might see,
> We thought he here had immortalitie.

An 'Epitaph' published alongside this 'Elegie' similarly declared:

> Is Atheist-Hobbes then dead! forbear to Cry;
> For, whilst he liv'd, he thought he could not dy,
> Or was at least most filthy loath to try.[171]

Hobbes had spent the years of the Civil War in France, where he studied medicine and befriended the philosopher René Descartes. Whether the latter's scheme came from Hobbes, from Bacon, from other influences, or from his own imagination, Descartes too believed that he could prolong his life for a long time – in fact, for a *very long time*. At the age of forty-one, Descartes had been disturbed to find that his hair was going grey. It brought a new focus into his life. As he told his friend Constantyn Huygens in 1637, from now on his principal subject of study would not be philosophy; it would be the search for a method of retarding the process of ageing.[172] He set his hopes high: the following year he told his friend that he hoped he might yet live 'more than a century' longer.[173] But he was soon scaling back his ambitions, telling another friend, Mersenne, in 1639, that he was expectant of at least another thirty years of life.[174] It was his good fortune, he explained, that in the last thirty years he had not 'experienced any illness worthy to be called an illness'. In fact, having acquired some knowledge of medicine and of how to take care of himself as if he were 'a wealthy gouty person, it seems as if I am now farther from death than I was in my youth'.[175] By 1645, he was informing Hobbes's patron, William Cavendish: 'The preservation of health has always been the principal end of my studies.'[176]

According to Descartes's seventeenth-century biographer, the philosopher followed a frugal diet (eating little, but often), drank sparingly, took moderate exercise, and kept a careful control of his emotions, having a love 'for peace and tranquility'. He never allowed his blood to be taken, and was careful in his choice of medicines.[177] All these things, Descartes believed, would help him prolong his life considerably. For if the body of man was a mere machine, a piece of divine clockwork, why should not the careful soul, through prudent diet, frequent exercise and careful repair, be capable of extending its operation and effecting its repair – indefinitely?

Like Bacon, Descartes believed that future ages would make immense progress in medicine. In his most famous philosophical tract, the *Discourse of a Method* (1649), he wrote that, 'if it be possible to finde any way of making men in the generall wiser, and more able than formerly they were,' it 'ought to be sought' through medicine. This science, he admitted, was still rudimentary, but 'whatsoever is known therein, is almost nothing in comparison of what remains to be known'.[178] One day, he suggested, 'we might be freed from very many diseases, as well of the body as of the mind, and even also perhaps from the weaknesses of old age, had we but knowledge enough of their

Causes, and of all the Remedies wherewith Nature hath furnished us'.[179]

Descartes concluded his celebrated work by formulating his resolution to 'employ the remainder' of his life 'in no other thing but the study to acquire some such knowledge of Nature as may furnish us with more certain rules in Physick than we hitherto have had'. Indeed, he explained that his 'inclination' for the importance of this project drove him 'strongly from all other kind of designes'.[180] Descartes believed that the only things that could inhibit this 'enquiry of so necessary a Science' were either 'the shortness of life' or 'the defect of experiment'.[181]

When the English philosopher and alchemist Sir Kenelm Digby met Descartes in Paris in the 1640s, they discussed '*la vie éternelle*'. Digby told him that, since 'life was almost too short to attain to the right knowledge' of things, he, Descartes, who so well understood the

Figure 1.4 René Descartes: detail from *La Reine Christine de Suède, entourée de sa cour*, by Louis Michel Dumesnil (Châteaux de Versailles et de Trianon). René Descartes's decision to go to Stockholm and teach philosophy to Queen Christina was the worst he ever made. On 10 April 1650 an Antwerp newspaper reported 'that in Sweden a fool has died who had claimed to be able to live as long as he liked'.

working of the human 'machine', ought to be searching out means to prolong its conservation. Descartes replied that he *had* considered this matter, and told Digby 'that to render a man immortal, was what he would not venture to promise, but that he was very sure it was possible to lengthen out his [i.e. man's] life to the period of the Patriarchs'.[182]

Such was Descartes's apparent confidence in his method of life extension that his last patron, Queen Christina of Sweden, had the impression he was seeking to live forever. But things did not work out so well. Christina had invited the famous Frenchman to Stockholm to teach her philosophy. He was reluctant to make a trip he judged to be dangerous to his health. Besides, he was busy working on a book devoted to medical physiology and anatomy.[183] But the Queen massaged his ego, and he agreed to go for three months. He arrived in winter, and unwisely acceded to the Queen's desire to take her lessons when her mind was properly 'disengaged from the incumbrance of affairs'. This, unfortunately for Descartes (who liked to work for long hours in the comfort of his bed), was five o'clock in the morning.

The cold quickly took its toll. Descartes ailed, was bled against his better judgement, and, in February 1650, died in Stockholm of pneumonia. He was only fifty-three. On hearing the news in London a few weeks later, the reformer Samuel Hartlib recorded in his diary that 'Cartes designe was to make a compleate Philosophy. In reference to this scope imagining that it was possible in nature to prolong ones life to a thousand years.' Hartlib – who had his own ideas of how human life could be prolonged – added that Descartes had chosen 'to live in the Low Countries as a free Commonwealth where hee might live as hee list and to follow his studies, fancying that hee might live a thousand years to perfect his Philosophy'.[184]

Descartes's friends were stunned by the philosopher's sudden death. The Abbé Picot declared that, if it had not been for that 'foreign and violent cause' which had 'deranged his "machine" in Sweden', Descartes 'would have lived for five hundred years'. An Antwerp news-paper was less impressed, reporting on 10 April 1650 'that in Sweden a fool has died who had claimed to be able to live as long as he liked'.[185]

With both Bacon and Descartes prematurely dead, this was not a promising start for the new science of prolonging life. But it was not the end of the seventeenth-century project. Whilst neither cold nor diet had worked, there was still another way.

CHAPTER 2

The Elixir of Life

In vain do Individuals hope for Immortality, or any patent from
Oblivion, in preservations below the Moon . . .

Sir Thomas Browne, *The Hydriotaphia, or Urn-Burial* (1669)

London, late summer, 1665 Even at the best of times the City is no
easy place to live. The 'Clowds of Smoake and Sulphur, so full of Stink
and Darknesse', of which John Evelyn has recently complained to the
King – made worse by overflowing gutters and narrow, labyrinthine
streets – render London more like 'the suburbs of the *Hell* than an
Assembly of Rational Creatures'. Bad air stunts the growth of plants,
and no fruits will grow on the trees of the polluted capital. 'Argument
sufficient', Evelyn suggests, 'to demonstrate, how prejudicial it is to the
Bodies of Men'.[1]

And now, how they die in droves! In recent weeks a law has been
passed, decreeing that the dead should be buried only by darkness. But
the dead are too many for this to have any practical meaning. The Bills
of Mortality for the last week of August record 7,496 deaths within the
City walls alone – 6,102 from plague. The following week the figures
will rise to 8,252 and 6,978 respectively. But informed citizens suspect
that even these are underestimates, that the true number might top
10,000 or even 12,000 dead from pestilence. The evidence is all around
them: corpses prostrate in the streets, coffins uncollected in yards and
side alleys, red crosses and prayers marked on doors, friends and

neighbours falling sick in the morning to be dead by twilight. Night and day, bodies are carried through the streets for burial in overflowing parish churchyards, to the doleful sound of passing bells. If ever John Donne's pulpit words rang most true, it's here in London in 1665: 'any man's death diminishes me, because I am involved in mankind, and therefore never send to know for whom the bell tolls; it tolls for thee.'[2]

Though two thirds of the shops in the fashionable Royal Exchange stand empty; though the College of Physicians has been abandoned and plundered by thieves, and the King and his court have decamped to Oxford; though the streets and river are 'mighty thin of people', London has not been entirely forsaken. Many, the diarist Samuel Pepys among them, continue about their business, but 'like people', he records, 'who had taken leave of the world'. Talk in the street 'is of death, and nothing else', he writes in his diary, 'this disease making us more cruel to one another than we are [to] dogs'. He carefully prepares his will, for 'the Sicknesse . . . is got endeed everywhere'.[3]

Whilst too many doctors have fled, some remain to tend the afflicted. Among them are the self-styled 'chymical physicians', followers of the more modern medical teachings of Paracelsus, Petrus Severinus and Jean Baptiste von Helmont. Though many have no formal university training, they have faith in their medicines and continue to serve their patients in this hour of need.

Among them is Dr George Thomson, a Royalist veteran of the Civil Wars, graduate of the universities of Edinburgh and Leiden. He visits the ailing and dying late into the night, enduring the abominable, noisome smells that erupt from plague carbuncles and the tainted, fetid expirations of decaying bodies. Though he curses 'the evil effects of *fugitive Physicians*', Thomson rejects the 'two principal Pillars' that support the classical Galenic system – bloodletting and purgation – both of which he can demonstrate 'to be pernicious to Humane Nature, destroying more than ever the Sword'. Instead, he uses modern medicines made from metals, minerals and remedying plants prepared in his laboratory – heated, purified, distilled by fire in his crude furnace.

His patients include Mr William Pick of Petticoat Lane, 'grievously wounded', as Thomson later records in his memoir of that dreadful year, 'with one of those poisonous Arrows that flew thick about poor

Figure 2.1 George Thomson: engraving from the frontispiece of his *Loimotomia, or, The Pest Anatomized: An Historical Account of the Dissection of a Pestilential Body*, 1666. The opportunity to conduct human autopsies was rare in seventeenth-century England. George Thomson (to the right) and his assistant are shown dissecting the pustulating corpse of a plague victim.

Mortalls'.[4] Thomson visits, dispenses his chemical remedies, and cures his patient. But one of Pick's servants, a young boy of fourteen or fifteen, is not so fortunate. He labours 'under most horrid symptoms, raving as it were extimulated by some Fury'. Too late to be saved by the doctor's medicines, the youth drops dead before him. Thomson asks Pick to grant him liberty 'to open this defunct body', and Pick consents.

Thomson is exhilarated – he has waited a long time for permission to dissect the corpse of a plague victim. He retrieves his surgical tools and later in the day returns to Petticoat Lane, where he finds the body laid out in a nearby yard. As seen in the illustration above, Thomson burns sulphur in a porringer beneath the open coffin and clears the burial clothes from the corpse: 'I could not but admire', he records in his autopsy report, 'to behold a skin so beset with spots black and blew, more remarkable for multitude and magnitude than any that I have yet seen.'

With his first cut he opens the bowels. A thin, yellow-greenish liquid oozes out. The small intestines are 'much distended with . . . a great quantity of a foul scoria or dross in them'. He is surprised that the corpse, though dead for some ten hours, remains extraordinarily warm. A dark, congealed substance encloses the inner vessels, whilst the liver weeps a thin, yellowish excrement. The spleen, dissected, appears 'more than ordinary obscure, a livid Ichorous matter following the Incision'. The kidneys, laid open, abound 'with a Citrine water', and the tender membranes of the stomach ooze a pint of black, ink-like matter, 'somewhat tenacious and slimy'. Cutting the vena cava and aorta, expectant of a considerable flow of blood, Thomson finds only a few spoonfuls 'of a thin liquor of a pale hew . . . which might easily be licked up by a small handkerchief'. The vessels are stuffed 'with a thick curdled blackish substance'. One ventricle of the heart offers up only a white unyielding matter, the other a black fuliginous substance, more fitting, he fears, for the infernal stomach of a hell-hound than for that of a human being.

Though he has not found the source of the sickness, he has seen first-hand its internal effects. His dissection complete, Thomson experiences a strange numbness in the hand 'which had been soaking and dabling in the Bowels and Entrals' of 'this loathsome Body'. Wiping off the 'foul gore', he holds his hand over a dish of burning sulphur. This, he will later realize, is a dangerous mistake. Instead of cleansing it, the sulphurous fumes dilate the pores of his skin and let in 'those slie, insinuating, venemous Atoms' of the plague. They will spread slowly now within him, 'like a Gangrene, diffusing its malignity into all my members'.

Unaware of what has happened, he returns to his lodgings, dines, and goes to bed. But at two o'clock that night he wakes. He has no sensation in his thighs or legs. The plague venom, he realizes, which had lain 'cryptick' within him, has been 'silently working in this subterranean microcosm' before acting 'publickly'. He realizes, with horror, that he has the plague.[5]

Immediately he directs his apprentice to prepare with all speed large quantities of his best medicaments, doubling and trebling the ordinary dose. Despite this, his head is still 'much out of order'. Small spots appear on his breast and arms – the disease is taking hold of his life-giving inner force, his Archeus. There is only one man in London who can save him now, and he sends for 'that Excellent expert Chymist and legitimate Physician, Dr. *Starkey*'.

Who is this Dr Starkey? Born in Bermuda to Scottish parents, and a graduate of Harvard College, Massachusetts, George Stirk – or Starkey as he is more commonly called – had arrived in London as a rich young man, fourteen years before. As self-styled 'philosopher by fire', he has spent his youth (as well as his wealth) in the pursuit of alchemical knowledge and the creation of brave new medicines. Like Thomson, he is a keen reader of the Flemish physician J. B. van Helmont's 1648 magnum opus, the posthumously published *Ortus medicinae*. The subtitle of the 1662 English translation of this immense tome had made van Helmont's intentions very clear: therein the 'whole Art' of medicine was to be 'Reformed & Rectified', providing 'a New Rise and Progress of Phylosophy and Medicine, for the Destruction of Diseases and Prolongation of Life'. In its 1,161 pages, van Helmont even claimed to have once possessed a tiny grain of the philosophers' stone, that miracle elixir which preserves both metals and living things 'from cankering, rust, rottenness, and Death and makes [them] to be as it were Immortal'.[6]

Starkey too is searching for this elusive medicine. On his arrival in London in 1651, it was rumoured that he had already 'done a number of most strange and desperate cures' in America, and he was soon teaching alchemy to no less a person than the Honourable Robert Boyle, one of the richest, brightest young men in Britain.[7] Starkey has told Boyle that he is close to establishing the recipe of 'an admirable medicine ... with a most desirable quicknesse & protractive of Old age Espetially'.[8] And he claims to know an 'Adept' in Massachusetts who possesses the secret of the philosophers' stone, who has used it to restore the hair and teeth of an old lady, and to make a withered peach tree bring forth new fruit. He professes to be on the verge of discovering the elixir himself – but it is an expensive, laborious, pious process. Some of his new friends in England – possibly members of Samuel Hartlib's intellectual circle – invest £300 in his alchemical processes, but with unfavourable commercial results.[9] When his grand projects come to naught, when he ends up in debtors' prison and struggling against his spendthrift ways and fondness for drink, Starkey's early supporters, discovering his 'rotten condition', turn their backs on him.[10]

In 1658 Starkey publishes a book, *Pyrotechny Asserted and Illustrated*, based on his laborious studies. It is dedicated to Boyle and describes a 'universal, and wonderfully restorative' elixir that cures 'all Diseases powerfully'. Though requiring 'tedious' effort to bring it 'to this height', Starkey considers this elixir to be 'a middle way' to preparing that greater medicine that will perform 'incredible cures', restoring 'the body of Man wonderfully, renewing really the Hair, Teeth, and Skin'. And he includes in his book another elixir, made from the oil of evergreen trees found 'in the cold frozen Regions' of Russia and Newfoundland, which 'doth also promise, at least gives hopes (on probable conjectures) of long Life, and of amending the Constitution, bringing it to a virile flourishing temper of Youth and Strength, and vindicating it from the cold declining Winter of wasting old Age'.[11] In another book, subtitled *A Short and Sure Way to a Long and Sound Life*, Starkey explains that nothing – aside from death itself – is beyond the curative powers of the 'true Physician' who 'knows that diseases are all in their kinde curable without exception'.[12]

The ailing Thomson has faith in Starkey. Though the Bermudan is also labouring under the pestilence, he comes to his friend's assistance. He brings with him a rare powder made from the corpse of a toad, and advises the doctor to hang a dried toad round his neck. Together, these powerful sympathetic medicines will draw the 'venemous Atoms' of the plague from Thomson's body: for six days he lies sweating on his bed. He attaches leeches to a bubo the size of a tennis ball in his anus: a 'filthy, blackish, ichorous, staining Recrement' issues from it, 'much like that which I found in the Cadaverous Dissection'.

But the remedies work. On the eighth day Thomson struggles downstairs and uses his own chemical remedies to cure his landlady, his maid and his apprentice: 'That Four in one House should have the Pest about the same time, and all escape when it raged most Mortally, is a Mercy never to be forgotten by us!', he subsequently rejoices.

Starkey, however, was not so fortunate. Falling foul of his old friend, drink, he had imbibed too much small beer when the plague finally struck him. The alcohol had subverted his body: his defences debilitated, his blood congealed, he knew, the moment he discovered a bubo in his groin, that none of his medicines, none of his elixirs, not even his mysterious, make-believe Massachusetts friend, could save him.

Within a few hours of visiting Thomson, George Starkey is dead. Their friends withhold this news from Thomson until they feel he has recovered enough to bear it. The doctor is devastated at the loss of this 'eminent light', this 'pillar' of chemical medicine who had devoted everything to fighting disease and to the search for the elixir of life. Starkey was only thirty-seven years old.

§

Alchemy, it seemed to many cognoscenti in seventeenth-century Europe and North America, promised both boundless health and, conceivably, immortality. The promise of incredible wealth – of creating gold and silver from base metals – was much less important. Some puritan alchemists even hoped that that discovery would render precious metals and money worthless, forcing men to focus on more important things – such as God. It was suggested by some that Adam himself had been an alchemist, and that he and the patriarchs had used the ultimate alchemical goal, the philosophers' stone, to prolong their lives almost to a thousand years. Rediscover the elixir of life, the alchemists thought, and death might be defeated. The loss of the likes of Starkey and other chemical physicians in the plague was a setback, but their passing did not – in the opinion of the alchemists, at least – discredit the validity of their mission.

Sir Francis Bacon had dismissed such assumptions. He saw the prolongation of life as a laborious task, not to be quickly won. Certainly, as he told the readers of *The History of Life and Death*, it could not 'be effected, by a few drops of *some precious Liquor*, or Quintessence'.[13] But that quintessence (the 'fifth element', the purist, most nourishing, medicinal essence of any substance),[14] the tincture, the philosophers' stone and the elixir of life were exactly what seventeenth-century alchemists such as Thomson, Starkey and Boyle *were* seeking. Nor was alchemy (or chymistry, as it was also called) some fringe feature of the Scientific Revolution. It occupied the minds of some leading philosophers of the age, most notably Robert Boyle and his young contemporary, the Cambridge natural philosopher Isaac Newton. Re-branded and stripped of its esoteric elements, it would be legitimized within the canon of the 'true' sciences in the eighteenth century, under the more innocent name of 'chemistry'.[15]

But what was alchemy? Why did it prove so popular that Charles II of England, Henry IV of France and Christian IV of Denmark all employed alchemists at their courts, and universities appointed professors in the subject?[16] Did they *really* believe it could open the route to prolonging human life for hundreds of years? As we shall see, the answer to this last question was an emphatic – if not unproblematic – 'yes'.

§

Alchemy's origins lie in obscurity. China, India and the Near East all raise claims to have been its first home. What is generally agreed is that it emerged in very ancient times, born from the rituals and wisdom surrounding the craft of metalworking.[17] Even in rudimentary metallurgy fire transforms lumpen ores into liquid metal. Out of the red hot furnace comes the purity of gold, silver and, most mysterious of all, quicksilver: when the bright red powder cinnabar is heated, almost miraculously it becomes mercury, the only metal which is liquid at room temperature. Heat it again, and it returns to a red powder. This seemingly magical cycle could be taken as a metaphor for creation: out of dust, dirt and fire springs forth life.

Alchemy thus included a strong religious element. Those who sought the secret truths of alchemy sought knowledge of the very essence of nature. Their ultimate goal was the philosophers' stone. This had the power to transmute base metals into that supposedly perfect and purest of all metals, one that never rusted: gold. But this elixir – this 'medicine' or 'panaceum', as it was also called – could cure disease, restore health and prolong life, perhaps indefinitely. In the third century BC, Qin Shi Huang, first emperor of China, sent servants in search of the elixir, and took mercurial medicines to achieve immortality. If anything, they shortened his life, and his body was buried among his famous army of terracotta warriors.

It was via the texts of Islamic scholars in Spain and North Africa that a knowledge of ancient near-eastern alchemical wisdom and traditions reached medieval Europe. The most famous of these Arabic scholars was Jabir-ibn-Hayyan – or Geber, as the Latin West knew him. Some two thousand texts are attributed to Geber, though the Brotherhood of Purity, an Isma'ilya sect, probably wrote most of them in the ninth and tenth centuries AD. The Islamic roots of Western chemistry can be seen in the way numerous associated English words carry the Arabic prefix '*al*', which simply means 'the': alcohol, alkali, alembic, elixir (deriving from *al-'iksir*, meaning 'the powder for drying wounds' or, perhaps, 'the immortality powder').[18] And then, of course, there is alchemy itself, a word coming through Greek: *al-khemia*, 'the art of transmuting metals'.[19]

These alchemical works were invariably obscurely written, using a panoply of metaphors and code-words to conceal their secret processes. Rather than putting off readers, however, such mysticism and secrecy appealed to the medieval and early modern love for intellectual riddles – as mottos, emblems, symbols and hidden iconography are to be found throughout the literature, art and architecture of these periods. These conceits emulated nature, the 'book' written by God and read (if not always understood) by people: nature, like the prophecies, parables and numerology of the Bible, was filled with wonderful, hidden knowledge – if only those secrets could be deciphered and comprehended.[20] As the Apocryphal Book of Ecclesiasticus promised: 'The Lord hath created medicines out of the earth; and he that is wise will not abhor them . . . with such doth he heal [men], and taketh away their pains.'[21] The doctrine of signatures was one attempt to interpret this hidden abundance.[22] Alchemy was another.

Alchemy, however, was a task of many lifetimes, and it supposedly involved a secret network of adepts who transmitted their knowledge across generations; only those wise, pious, committed, trustworthy and long-lived enough could ever hope to fathom its ultimate goal and to come to know the hidden workings of creation. The challenge proved irresistible to numerous early English scholars – Elias Ashmole would claim that 'no nation hath written more, or better', on alchemical subjects than the English.[23] It was an Englishman named Robert of Chester who in 1144 completed the earliest known translation of an Arabic alchemical work into Latin. A century later, the Oxford scholar Roger Bacon included an extensive discussion of alchemy in his *Opus tertium* (1267). He also wrote a long tract on how human life might be prolonged. Translated into English in 1683, his book recommended medicines made from snakes, to repair 'the Faculties and Senses' and restore 'the Strength of the Body'. (Snakes were considered a valuable medicament because of their ability to shed their skin and seemingly remain eternally young.) Another medicine, made from a herb 'like Marjoram', if taken for several days with cow's milk, 'the Grey hairs will shed, and Black ones come in their room, and the Man will become more juvenile'. Bacon also noted that wine was a good health-preserver, for it 'cheers the Heart, tinges the Countenance with Red, makes the Tongue voluble, begets Assurance, and promises much Good and Profit'.[24] Roger Bacon's Spanish contemporary, Arnald of Villanova (1235–1311), was also interested in alchemy and, like Bacon, pondered over its medicinal potential. Many alchemical medicines were to be made from metals such as mercury, whilst others depended upon distillation, a key operation in the alchemist's laboratory.

Though many alchemists were more concerned with the medicinal aspect of their art, to the rulers of Europe their alleged ability to make silver and gold was equally alluring. Whilst Henry IV outlawed alchemy in England in 1403, Henry VI hoped that alchemists might solve his financial problems. However, in 1456 he granted letters patent to three men who petitioned him for permission to search for the elixir of life. This was 'a medicine', they explained to the king, 'whose virtue would be so efficacious and admirable that all curable infirmities would be easily cured by it; human life would be prolonged to its natural term, and man would be marvelously sustained unto the same term in health and natural virility of body and mind'. One of the three men awarded the patent was the king's own physician.[25] But Pope John XXII's condemnation of 1317 – 'They promise that which they do not produce' – haunted alchemists down the centuries.[26] Theirs, it appeared to many critics, was a suspicious, specious, and even perhaps an heretical claim to knowledge.

Alchemy enjoyed a revival in the Renaissance, proving to be of interest to a number of prominent Italian scholars. But its greatest proponent was the radical sixteenth-century German physician Philippus Aureolus Theophrastus Bombastus von Hohenheim (c.1493–1541), better (and more conveniently) known to history as Paracelsus. Son of a physician, Paracelsus's early life is obscure. He spent time working in mining, but from the age of fourteen wandered Europe. It is possible that he visited the Near East; possibly, too, he obtained an MD from an Italian university. His earliest published works – of which there are many – were on religious prophecy; for this was the height of the

Reformation, and religion permeated all aspects of society. His first medical publications were two short tracts on syphilis, a new disease, apparently beyond the control of the medical profession. Paracelsus rejected many traditional preventative and curative methods based on the writings of ancient doctors and philosophers; he is alleged, for instance, to have burnt in public a copy of Avicenna's works: modern times, he explained, demanded modern means. Contemporaries labelled him the Martin Luther of medicine. Paracelsus updated the alchemy of the Middle Ages, relaunching it as the new route to curing disease and prolonging life.[27]

Among Paracelsus's many medical works, two are of particular interest: *A Treatise Concerning Long Life* and *The Book Concerning Renovation and Restauration*. In his introduction to the former, Paracelsus asserted that it was 'a most certain truth' that the body 'may be restored, changed to the better' and 'wholly renewed', and life prolonged. Clearly, if a dead body could be long preserved 'from putrefaction or decay' by embalming, 'how much more may a living Body be so preserved'.[28] Paracelsus likened life to 'a thing spiritual, like hearing, seeing, smelling'. Fire does not exist in steel or flint, yet when the two are struck together, they create a spark, which creates fire: it is the same with the body and life, for 'life comes from these things, in which there is not life'. But the question of how 'any thing can come from that in which it never was' cannot be answered. It is enough to say that life is like fire, and the body is the fuel that feeds its flames:

> We ought not then to think the time of mans life is so determined, that every man must needs die such a day and such an hour; nor is it suitable to Christianity, to think this, that the life cannot be prolonged by these Medicines which God hath created to this purpose: God hath created these Medicines for our use, and he hath granted us a liberty to use them; the defect is onely in us, that we do not know those Medicines.[29]

For Paracelsus, Adam had known all these things 'perfectly'. If he had not been 'deprived' of access to the Tree of Life he would have 'continued immortal'. The Tree of Life was, therefore, not just the subject of biblical studies. It was 'a natural Tree', and had a part to play in medical theory.[30]

Figure 2.2 Paracelsus: line engraving, anonymous, sixteenth century. The death of Paracelsus before he was even fifty undermined his claim to possess secret medicines that could prolong life well beyond a hundred years. Nonetheless, he and his followers launched a revolution in early modern medicine that helped to overturn the ancient Galenic system.

Like the ancient physicians, Paracelsus fully acknowledged that the places where we live, together with the things we eat and drink, play important roles in prolonging (or shortening) life. Meat and drink, 'if rightly used', preserve it; but, if 'abused', destroy it.[31] Indeed, he wrote that three things were to be observed in 'preserving the Body': diet, disposition and medicine. But, out of these three, unlike Bacon or Descartes, Paracelsus considered medicine to be 'the chief thing'. Medicine alone could preserve man 'however it finds him, yong or old' – but only those with the right understanding would be able to understand 'these remedies of long life'.

Paracelsus wrote of various medicines that would 'renovate' and 'restore' the human body. They included both 'simples' and quintessences: some were capable of preserving life for forty years, others, of prolonging it for a hundred or more. Hellebore, he asserted, 'can preserve life an hundred and twenty years', if rightly used.[32] The medicines recommended in *The Book Concerning Renovation and Restauration* included the 'first entity' (*primum ens*), or the inner, active 'virtue' or power of minerals and plants. These 'virtues' could only be extracted by chemical means. The 'tincture', 'arcanum' or quintessence of a substance was its inner essence (*ens*). This, as the sixteenth-century Swiss botanist Conrad Gesner explained, was 'the chief and the heavenliest power or vertue in any plant, metal, beast, or in the partes thereof, which by the force and purities of the hoale substaunce . . . conserveth the good health of mans body, prolongeth a mans youthe, differeth age, and putteth away all manner of diseases'.[33] The cures claimed by the Paracelsians for such chemically contrived arcana were remarkable. For Paracelsus, there was no predetermined length of human life, so long as the right medicine – be it a quintessence, an elixir or the philosopher's stone – could be found to cure disease and maintain health.

The physical effects of these medicines were sometimes dramatic, however, and it is obvious why many traditionally minded physicians opposed their use. In his *Book of Renovation and Restauration*, Paracelsus wrote that his 'renovating' medicines, which were made from the 'first entity' of antimony, sulphur, gold and 'herbs' such as opium, were to be taken daily 'so long, till your nails of your fingers first fall off, and then the nails of your feet, then your hair and teeth; and then lastly, till your skin be dried up, and new bee again generated'. Only then was the medicine to be discontinued; in due course 'so will there new nails be born again, new hairs, new teeth, and withal, a new skin; & the diseases both of the body and mind will depart away'.[34]

Early modern chemists sought to emulate these effects. In the 1650s John Evans, an impoverished Oxford graduate and disgraced former minister, explained how an 'ancient man' of eighty-four had used his medicinal antimonial cup, to treat scabs covering his whole body. Within a few days, Evans declared, 'his hard crusty scabbedness scaled off, and his flesh and skin was restored and renewed both white and cleare like a young mans'.[35] In the 1660s the self-styled 'Unlearned Alchymist' Richard Mathew of London prescribed a chemical pill to a gentleman presumably suffering from syphilis, with similar effects. Mathew was both startled and impressed when his patient returned after some three weeks of taking his medication, 'and shewed me his naked body, which I was loath he should, and [there was] not one hair upon him, but a fresh skin, as of a young child'. The gentleman told

Mathew that 'he was as well as ever he was in all his life', and what made Mathew 'more to wonder, was that the nails of his hands did then begin to peep out, like the little white that is at the root of our nails'.[36]

Mathew also claimed that, 'although to many it may seem incredible', it was reported to him by another gentleman that an old lady 'aged betwixt eighty and ninety' had taken Mathew's pills 'for some years' and 'she now hath young teeth growing in her head'. Her periods had also returned 'as when she was but 20 years old'.[37] If these accounts were true, they were probably the pernicious effects of poison: arsenic, mercury and antimony were common ingredients in chemical medicines. This was an early modern equivalent of chemotherapy, and such symptoms were taken as the clear and welcome signs of rejuvenation.[38]

The medicines certainly appear to have been popular. Though to many traditional physicians such powerful concoctions killed as often as they cured, belief in their efficacy had gained widespread support as early as the 1560s. John Thornborough, Bishop of Bristol and Worcester, was 'much commended' for his 'skill in chemistry', and was said to have presented 'a precious extraction' to King James I which was 'reputed a great preserver of health, and prolonger of life'. When the bishop died 'exceeding aged', in 1641, he was at least in his late eighties.[39] The Huguenot apothecary Gideon Delaune, who emigrated to London around 1572, was, to judge by his will and a funeral certificate from the College of Arms, aged ninety-four when he died in 1659.[40] According to contemporaries, Delaune had made thousands of pounds from a proprietary pill he manufactured; given that one critic stated that it was 'not a very safe one', it may well have been a chemical medicine.[41] It appeared that at least *some* men were able to use them to good effect in prolonging their own lives.

§

Whilst Paracelsian ideas influenced Sir Francis Bacon, the latter had no time for what he called 'impostors' in philosophy.[42] As early as 1610, in Ben Jonson's eponymous play the alchemist was suspected of charlatanism. 'In eight, and twenty dayes', Jonson's impostor promises, 'I'll make an Old man, of fourscore, a Childe . . . Restore his yeares, renew him.'[43] Though Bacon conceded in *Novum Organum* that alchemists had 'made several discoveries', and (albeit accidentally) 'presented mankind with useful inventions',[44] it was undoubtedly alchemists he was attacking when he noted the 'many silly and fantastical fellows who, from credulity or imposture', had 'loaded mankind with promises, announcing and boasting of the prolongation of life, the retarding of old age, the alleviation of pains, the remedying of natural defects, the deception of the senses, the restraint and excitement of the passions, the illumination and exaltation of the intellectual faculties, [and] the transmutation of substances'.[45]

It is also possible that Bacon was ridiculing the so-called Brother-hood of the Rosy-Cross, which had first emerged in the 1610s. The Rosicrucians claimed to possess the secret of – among other rare knowledge – the philosopher's stone. 'Were it not a precious thing', the mythical Christian Rosencreutz asked in the *Confessio Fraternitatis* of 1615, 'that you could always live so, as if you had lived from the beginning of the world, and, moreover, as you should live to the end thereof?'[46] In 1623, posters appeared in Paris declaring that Rosicrucians, both 'visible and invisible', were abroad in the city, 'in order that we may rescue our fellow men from the error of death'.[47]

Who was, and who was not, a Rosicrucian, no one quite knew. Their writings seemed to carry more weight than substance, but their influence and secret presence lingered. In 1661 a prominent London publisher claimed that a man purporting to be one of only twelve Rosicrucians in the world had offered him a manuscript that explained how to 'Cure all Diseases, make old Persons become yong again, [and] Restore the body to such youth and strength that a man should live some hundreds of years longer'.[48]

Potable gold and other such 'chymicall medicines' received short shrift in Bacon's *History of Life and Death*. They 'first puffe up with vain hopes', he complained, 'and then faile their Admirers'.[49] As he wrote in *The Advancement of Learning*, only someone who had studied 'perfectly' the processes of the human body, and who had investigated thoroughly the effects of diets, baths, ointments and 'proper Medicines', would be able to prolong their life – or at the least 'renew some degrees of youth, or vivacity'. Notwithstanding his reservations about alchemists, however, it is notable that, whilst in *The Advancement of Learning* Bacon ridiculed the idea that silver could be turned quickly into gold by 'a few graines of *Elixir*', he did not doubt 'that

Gold by an industrious and curious wit, may, at last, be produced'.[50] Nor did he deny in *The History of Life and Death* that gold could be used 'to make good medicine' (though he did ridicule those who used such ingredients with 'a proud hope of Immortalitie').[51] Furthermore, in *Sylva Sylvarum*, Bacon observed that, though '[t]he World hath been much abused by the Opinion of *Making of Gold*: The *Worke* it selfe I judge to be possible'. He even presented an 'Experiment Solitary', suggesting how it might be done.[52]

Bacon's real objection to the alchemists was not their belief in transmutation or in the medical potential of gold. It was the fact that their 'practice' was 'full of Errour and Imposture', and their theory 'full of unsound Imaginations' – including their beliefs in astrology, natural magic and superstition.[53] These were ideas that failed to stand up to the rigours of Bacon's own legalistic intellectual method.

Others were equally critical. The well travelled Scottish physician Dr James Hart observed in 1633 that it 'cannot be denied, that by vertue of a laudable diet, that life of man may be prolonged to an hundred, or an hundred and twenty yeeres'; but that it might be perpetuated forever, he declared, 'is impossible'. So, he asked, 'what is the ordinary period whereunto the life of man by meanes of art may be prolonged?'

> Our ordinary Authours, as wee have said, assigne 100 or 120 [years]: but wee have a certaine sort of people, who in shew, would seeme to transcend vulgar understanding, and tell us strange things of the prolongation of mans life for many yeeres, farre beyond this above-mentioned period; and that by meanes of certaine medicines made of metals, of gold especially; and these be *Paracelsus* and his followers . . .[54]

Hart utterly dismissed such claims: 'is it not a thing ridiculous, now in these later times', he asked, 'to extend the life of man-kinde to 1000, 900, or at the least to 600 yeeres?'[55] In a crushing rejoinder, he pointed out that, notwithstanding the great merits of his supposed medicines, Paracelsus had died before he was even sixty years old.[56]

In actual fact, Paracelsus had died before he was even *fifty*. This was a potential embarrassment to his followers, seriously undermining their prophet's claim to have possessed life-extending medicines. Though

Jan Baptiste van Helmont was a keen supporter of that 'good man' Paracelsus, from whose writings he declared he had 'profited much', he considered him 'no less ignorant of a Medicine for Long Life, and the use thereof, than of the very Essence and Properties of Long Life'. So, whilst van Helmont held Paracelsus's arcana to be good medicines for 'healthy or sound Life, or unto a removal of Impurities; yet they do not any thing directly and primarily tend to long Life'.[57] Van Helmont, therefore, believed that it was necessary to look elsewhere for the true methods of prolonging life, the potential for which he appears to have measured by the hundreds of years.

Like the ancients, van Helmont first advised his readers to take care of their diet and surroundings, noting that there were certain dry, mountainous places 'at this day, whereunto a Life of three hundred years is ordinary'.[58] Those who lived cheerfully, 'far from the cares, usuries, busie affaires, and stormes of their age', were likeliest to live longest, and he advised his readers to avoid 'carnal Lust', gluttony, drunkenness, tobacco, frequent baths, bloodlettings, 'loosening medicines', and to live away from bad climates and contagious air.[59]

In all these respects, van Helmont's guidelines for long life were little different from those famously advocated by Luigi Cornaro in his *Discorsi della vita sobria*.[60] But, like Paracelsus and Bacon, van Helmont went further than simply advocating a 'regular' lifestyle. The Tree of Life that grew in the Garden of Eden, and which had promised eternal life to Adam and Eve, was van Helmont's medicine of choice for indefinitely prolonging life. It was also his arbiter for what could be achieved through nature's bounty. Whilst the Paracelsian *Arcana* could *cure* diseases, van Helmont believed that only the Tree of Life could *preserve* and *renew*, 'making young again . . . the vital Faculties'.[61] Though he had searched for it, van Helmont did not hold out that mortal man would ever rediscover this tree. Nor did he mention the fabled fountains of youth, described in classical works and supposedly sought among the Caribbean islands by Christopher Columbus's colleague, Juan de Ponce de León, in the 1510s.[62] Rather, van Helmont sought something *akin* to the Tree of Life – something with similar, if not actually identical, restorative properties.

He could not find them, however, in plants or animals. Snake's flesh and hart's horn were considered key ingredients in rejuvenating medicines, but 'even these things die', he observed, 'and are dead when

we use them for medicines, so how can they convey the spirit of life?'
He looked to gems and the philosopher's stone. But

> neither have I found in those, the Foot-steps of long Life . . . because
> they cannot be immediately assimilated, or adjoined unto our first
> constitutive Parts . . . At length, I concluded with my self, that
> whatsoever it were that should supply the Place of the Tree of Life,
> it was the Young or Off-spring of a Tree: And then, that this
> Medicine was to be fetched out of a most wholesom, odoriferous,
> balsamical, and almost immortal Shrub . . .[63]

Van Helmont believed that the likeliest candidate for this 'immortal
Shrub' was the 'Cedar in *Libanus*' or the 'Wood *Cetim*', from which
Noah had built the Ark. But it was not enough, he explained, simply
to use the fruit, bark, leaves, or sap of this tree from the shores of
Palestine.[64] The preparation of this medicine 'is the most exceeding
difficult of all those things which fall under the Labour of Wisdom'.
Van Helmont's method depended upon distilling the wood in a sealed
glass vessel for many months with an equal measure of what he called
the 'immortal solvent' – that is, the so-called 'Liquor Alkahest'.[65] And
here was the rub. What, exactly, was the Liquor Alkahest – and where
was it to be found?

As Paulo Porto explains, the Liquor Alkahest 'was an important
means for preparing medicines and for unveiling some of the deepest
secrets hidden in natural bodies. . . . only through the alkahest would
the physician be able to cure hitherto "incurable" diseases, and to
prepare a medicine for prolonging human life'.[66] It was Van Helmont
who first developed the idea of the Liquor Alkahest fully (from a hint
he found in reading Paracelsus), and the Dutch chemist Johann
Rudolph Glauber (1603–70), the first who saw it as the key to
discovering a range of remarkable medicines. Both men would be
enormously influential on the pursuit of chemistry and the search for
chemical medicines from the 1640s until the end of the century; the
Alkahest, like the elixir, became an elusive goal sought by later
seventeenth-century chemists.

§

It is through the extensive letters and diaries of the utopian reformer Samuel Hartlib (c.1600–62) that we learn most about the seventeenth-century search for the Liquor Alkahest and the philosophers' stone. Born to an English mother and a German father in the Baltic port of Elbing, Hartlib studied at Cambridge University, settling permanently in England in 1628. Over the following decades he cultivated an international network of correspondents, and in his diary, the 'Ephemerides', documented what he heard on a broad range of scientific and philosophical subjects. (It was there that Hartlib recorded how Descartes had fancied 'that hee might live a thousand years to perfect his Philosophy'.)[67]

Hartlib's colleagues in England included the alchemist Thomas Henshaw (1618–1700). Educated at University College, Oxford, Henshaw formed a circle of chemically inclined friends which included the author and diarist John Evelyn, the physician Dr Robert Child (1613–54), and the astrologer and antiquary Elias Ashmole (1617–92). Hartlib records how in 1650 Henshaw and Child were among those 'endeavouring to forme a Chymical Club' with the purpose of making 'all Philosophers acquainted one with another and to oblige them to mutual communications'.[68] One result of this collaboration may have been Ashmole's *Theatrum Chemicum Britannicum* (1652).[69] In his 'Prolegomena', Ashmole identified four grades of philosopher's stone, of which the fourth, the 'Angellical Stone' or 'Food of Angels', was the most powerful: it was, he wrote, the '*Giver of Years*, for by it *Mans Body* is preserved from *Corruption*, being thereby inabled to *live* a long time without *Foode*: nay 'tis made a question whether any *Man* can *Dye* that uses it'. With these four stones, Ashmole explained, 'In Briefe, by the true and various use of the *Philosophers Prima materia* . . . the perfection of *Liberall Sciences* are made known, the whole Wisdome of Nature may be grasped'.[70] He believed that at least some of these stones were known to men: in 1653 he recorded that the adept William Backhouse, being ill and thinking he was about to die, revealed to him 'the true Matter of the Philosophers Stone'.[71]

A young gentleman who joined Hartlib's circle in the late 1640s was the Honourable Robert Boyle. Born in Ireland in 1627, Boyle was the youngest son of a remarkable self-made man. Like Sir Nicholas Bacon, Richard Boyle rose from obscurity to high office: in his case, as 1st Earl of Cork and Lord High Treasurer of Ireland. His son Robert was educated at Eton College, and was in Geneva on the Grand Tour when his father died suddenly in 1643: the money Robert inherited was enough to set him up for life. On returning to England he moved into the estate his father had bequeathed him at Stalbridge, Dorset, occupying himself first as an author, then as an alchemist. In due course he became the leading chemical philosopher of the age.

Of slight build, Boyle was described by his young friend John Aubrey as being 'so delicate' as to compare 'to a chrystal, or Venice glass; which, though wrought never so thin and fine, being carefully set up, would outlast the hardier metals of daily use'.[72] He was also hypochondriac. In 1653 a friend warned: 'The . . . disease you labour under is your apprehension of many diseases, and a continual fear that you are always inclining or falling into one or another.'[73] It was almost as much through wanting to cure himself as it was through wanting charitably to help others that Boyle displayed a life-long interest in medicine.

Boyle's early ambition was to be an author. A deeply religious man who never married, he began his career penning ethical essays. But, after installing a laboratory at Stalbridge, and with George Starkey's initial guidance in the art of chemistry, Boyle's first published work appeared in a 1655 collection of essays titled *Chymical, Medicinal and Chyrugical Addresses*, dedicated to Hartlib. In his anonymous contribution, titled 'Invitation to a free and generous communication of secrets and receits in physick', Boyle chided anyone who possessed the elixir but kept it hidden. He pointed out that it was a discovery 'that we owe wholly to our Makers Revelation, not our own industry . . . And therefore though God should address those special favours but to some single person; yet he intendeth them for the good of all Mankind'.[74] At least some contributors to the *Addresses* had high hopes for imminent great discoveries in chemistry: an appendix portended 'that within a short time we shall have an universal Medicine, which will not onely recover the sick and keep them well, but also take away death, and for ever swallow it up'.[75]

In the early 1650s Boyle and Starkey earnestly pored over van Helmont's *Works* together, and were inspired by what they read.[76] Boyle acknowledged that he had 'no mean esteem of divers Chymical Remedies'[77] and that he was for all his life a searcher after the philosopher's stone. Though he later questioned certain chemical

Figure 2.3 Robert Boyle: engraving of 'the Honourable Robert Boyle' by George Vertue, after Kerseboom, 1739. Robert Boyle spent much of his life in search of the secrets of nature through the ancient and mysterious art of alchemy: the philosopher's stone promised to reveal a powerful medicine or 'elixir' that might cure all human ills – including death.

claims and methods in one of his most famous work, *The Sceptical Chymist* (1661), Boyle's reservations were not so strong as the title of that book has led some to assume. He never dismissed the possibility that the philosopher's stone existed, and to the end of his life corresponded with chemists throughout Europe who were seeking it.

Indeed, in a short, anonymously published essay of 1678, *Of a Degradation of Gold Made by an Anti-Elixir: A Strange Chymical Narrative*, Boyle gave an account of an 'Experiment' he had made with a tiny quantity of what he called, variously, '*Anti-Elixir*', '*Anti-Philosophers Stone*' or 'Medicine', which he had obtained from a stranger who had travelled to the East. In his essay, Boyle recounted his experiment as a 'matter of Fact' presented before 'an Assembly of *Philosophers* and *Virtuosi*' which was headed by a 'President' – terminology clearly suggestive of the Royal Society. A dark reddish powder, Boyle claimed, had transmuted molten gold into a lesser, silver-like metal. Given the apparent success of this 'anti-elixir', Boyle asserted: 'I see not why it should be thought impossible that Art may also make a *true Elixir*.'[78] In the following century, this essay would be described as having made 'a great noise both at home and abroad', and as being 'one of the most remarkable pieces that ever fell from his pen; since the facts contained in it would have been esteemed incredible, if they had been related by a man of less integrity and worth than Mr. Boyle'.[79]

It did not seem improbable to Boyle, therefore, that an elixir – whose effects would include the prolongation of human life – existed *somewhere* in nature: it was simply awaiting discovery by someone diligent and pious enough to be rewarded by God for his (or indeed her) labour. Some, perhaps, had already been fortunate enough to find it: in 1680 Boyle publicly acknowledged his belief that there might live 'conceal'd in the world, a sett of *Spagyrists* of a much higher order than those that are wont to Write courses of *Chymistry*'. If they existed, these *Adepti*, Boyle believed, would possess the secret of the Alkahest, and would be 'able to transmute baser Metalls into perfect ones, and do some other things, that the generality of Chymists confess to be extreamly difficult'.[80]

Boyle appears to have sincerely believed the prolongation of life was possible. In a work-diary from the last years of his life he even recorded how an unnamed 'person' who had recently performed 'some extraordinary things in Chymistrie' told him that he had known in Venice 'an excellent Artist' – that is, one adept at alchemy. According to Boyle's informant, this Venetian had claimed that though he 'seemed to be at most between 40 & 50 year old yet in reality he was then 173 years of age'. Boyle noted that, although this story seemed 'scarce credible', he was 'less disposd' to dismiss it, because the person who had told it to him appeared to be 'noe Charleton but a plain honest German of good repute' from among some of Boyle's friends. Furthermore, from Boyle's other conversations with this man, he 'seemd carefull not to affirm things that he had not tryed or did not otherwise know to be true; nor did hee at all pretend to bee acquainted with any of this Artists secrets for the prolongation of life'.[81]

His preparedness cautiously to accept this account clearly depended in part upon the good character of Boyle's informant. But it must also have been influenced both by his – and his contemporaries' – search for the elixir, and by a wider cultural belief that human lifespans *could* be thus extended. It was a remarkable suggestion. But it did seem possible – to Boyle, at least – that it was true.

§

King Charles II of England was fascinated by alchemy. On his Restoration in 1660, he invited the devoted French Paracelsian Nicaise Le Fèvre to England, providing him with a laboratory at St James's Palace. According to Le Fèvre – who in December 1661 became a Fellow of the Royal Society – in the spring of 1663 Charles ordered him to apply himself wholly to the preparation of the famous 'cordial' which had been invented earlier in the century by another great chemist, Sir Walter Raleigh. Raleigh's Cordial incorporated everything considered good for the preservation of health and prolongation of life. Ingredients included hart's horn (because 'there are but few Animals that can equal the Hart for length of life, since he lives whole Ages') and gold ('because it re-establishes and augments the radical Moisture and the natural Heat').[82] At the suggestion of the chemist Sir Kenelm Digby and of the King's chief physician, Sir Alexander Fraiser, Le Fèvre added 'the Flesh, the Heart, and the Liver of *Vipers*' to Raleigh's recipe, because the snake renews its skin annually, and so 'the remedy it yields may also produce in us Renewing Principles and Faculties'.[83]

Raleigh's Cordial would be among the official medicines recommended during the Great Plague of 1665; John Aubrey later recorded that 'Mr. Robert Boyle haz the recipe, and makes it and does great cures by it'.[84] Boyle certainly reported in 1663 that this restorative medicine 'has been abundantly recommended by Experience', and that it had achieved 'many remarkable (and some of them stupendous) Cures'. Those he knew to have used it with success included an unnamed knight, 'who after all that Skilful Physitians could do, had long lain a dying'. Boyle recommended (or even supplied) Raleigh's Cordial to the dying man. Having heard no more about it, Boyle chanced 'to meet this Knight at *White-hall*, well, lively, and with a Face whose Ruddiness argued a perfect Recovery, and yet he is not very farre from seventy Years of Age'.[85] Perhaps as a reflection of his success as an unofficial physician, in September 1665 – only a few weeks after Starkey's death – Oxford University conferred on Boyle the honorary degree of doctor of medicine. Charles II himself would be given Raleigh's Cordial during his last illness in 1685 – though without success.[86]

Boyle's intellectual interests went far beyond those of Starkey, Raleigh or Le Fèvre, however. As it had been for Bacon, so for Boyle, too, almost the entire natural realm lay within his remit of interest. Around 1655 Boyle moved to Oxford, where he became closely involved in one of England's leading philosophical circles. He studied the latest writings on atomist theory and embarked on a remarkable programme of research and writing in natural philosophy. His work in Oxford included that for which he is probably best known today: his development, together with Robert Hooke, of a vacuum pump, and their experiments in physics that gave us both Boyle's and Hooke's Laws. But a great deal of his studies revolved – if not explicitly, then often implicitly – around the fundamental questions of the nature and operation of life.

In *The Origine of Formes and Qualities* (1666), Boyle observed that 'life' was 'a word whose meaning is not yet defined', and one that was 'apply'd to subjects that are exceeding different'. Some philosophers, for example, believed that gems and minerals were alive, since it was commonly supposed that they grow in the earth, and growth could be included in a definition of life.[87] Naturalists, meanwhile, argued as to whether hibernating insects were alive or not:

> not onely because those Insects seem to be devoid of sense and motion, but because they place the notion of life in a constant Circulation of the blood or some analogous Juice . . . whilst others on the contrary think them to be rather Benumm'd then Dead, because regularly recovering the manifest actions of life in the Spring, (or oftentimes before, if a due application of heat be made unto them) it cannot be suppos'd that they were during the Winter really destitute of Life: Death being a Privation, which by Physical meanes admites not of a return to the former state . . .[88]

But if death is to be defined as the absence of motion, what was to be made of physical activity in bodies *post mortem*? In *The Origine of Formes*, Boyle pondered over the expulsion of excrement from corpses and the suggestion that hair and fingernails continue to grow. He noted that the sixteenth-century French surgeon Ambroise Paré had embalmed and preserved a human body for twenty-five years, and that it 'still remain'd whole and sound, and that, as to the nailes, he found that *having often par'd them, he still observ'd them to grow again to their former bignesse*'.[89]

Hearing that children, 'for sport', revived drowned flies by warming them, Boyle tried the experiment himself, using wasps and bees: 'and having drown'd them so, that if let alone they would not in probability have ever recover'd', he found 'that the heat of the Sun would recover them as well as it has been observ'd that warm ashes would recover Flies . . . and the degrees and manner of their recovering again the operations of life suggested observations not unworthy to be taken notice of'.[90] Like Bacon, Boyle offered directions for further study – but, as in this instance, did not always pursue them himself. They were directions for future projectors, hints towards new theories of life and death.

As well as reviving wasps and bees, Boyle put live animals into a glass chamber and then extracted the air using his vacuum pump. The animals died. In the same experiment, using a lighted candle, he observed that the candle would go out. This demonstrated that there was clearly something in air supportive of both fire and life – but what exactly remained elusive. Nevertheless, the experiment added weight to the theory Boyle and certain contemporaries were exploring that matter consists of tiny atoms, or corpuscles, and that their interaction explained sensory and motive effects. Man was certainly a *machine*, but not one that operated by cogs, weights and pulleys alone: the flow of liquids, the movement of subtile gasses (a word invented by van Helmont), and heat-producing processes similar to fermentation, clearly all appeared to play a part in that mysterious operation known as life. Increasingly, life was to be defined as a *chemical* (in its broadest sense) rather than simply a *mechanical* process, and it was these processes, somehow gradually wearing themselves out, that ultimately caused our demise. Life and death were thus more complex processes than in the Aristotelian–Galenic analogy of a flame burning up a candle. The historian Barbara Kaplan suggests that, according to Boyle's theory of life as laid out in *The Origin of Formes and Qualities*, '[t]he reason the living being does not last forever is that nature impinges and motions inevitably are altered, thus changing the "texture" of the organism to one more difficult to be maintained'.[91]

Could this somehow be linked to the elixir's power vastly to prolong life? Boyle does not appear to have made this connection. But his interests carried him further than heat, cold and gasses. In another of the various thoughts for future investigation presented in *The History of Life and Death,* Sir Francis Bacon had pondered if there was some way to convey 'a young Mans *Spirit*... into an Old Mans Body'. If this could be achieved, Bacon reckoned, it was 'not unlikely, but this great Wheele of the *Spirits*, might turne about the lesser wheele of the Parts', and 'so the Course of Nature' could be turned backwards. But how this might be done, he did not know.[92] Whether this statement inspired them is not known, but between 1657 and 1665 Boyle and some of his Oxford colleagues – among them the astronomer and architect Christopher Wren and the physician Richard Lower – conducted the first transfusion of blood between live animals. Then in 1667 Jean Denis, Professor of Philosophy and Mathematics in Paris, made some seemingly successful transfusions of lamb's blood into human subjects. These were done with the express intention of 'curing sundry disease' and in the belief that 'hot blood' could 'reinfuse new strength into that which languisheth with coldness', and thus rejuvenate the prematurely aged.[93] Denis noted that it had been suggested that the operation be done using human blood, perhaps taken from a condemned criminal, but he believed 'it would be a very barbarous Operation, to prolong the life of some, by abridging that of others'.[94]

The Royal Society carried out the same experiment later that year. They used as their paid volunteer an impoverished young Cambridge graduate named Arthur Coga. Described as 'a very freakish and extravagant man', on 23 November 1667 Coga received arterial blood from a lamb through a cut in his arm. It was hoped that the transfusion might cure his mental instability, and within a few days Coga's landlord described his lodger as 'more composed than he had been before'. Coga himself declared before the Royal Society that he was 'as a new man'. But following a second transfusion it was found that 'the wildness' of Coga's mind remained 'unchanged'. The fact that he had spent the fee he had received from the society on alcohol was not thought to have helped his condition.[95] The operation, it would appear, had not been a success.

Nevertheless, the following year, Timothy Clarke told the society's secretary Henry Oldenburg of his hope that such experiments might cure diseases, or even 'restore in some measure youthful vigour to senility'.[96] But the experiment was not repeated. A subject who had received calf's blood in a similar experiment in Paris had died, and the procedure was effectively banned in France. Blood transfusion as a method for revivifying the aged lay dormant for almost a century and a half.

Robert Boyle, however, sought an alternative way of harnessing the power of blood. He suggested that a medicine, 'perhaps nobler, than the most costly and elaborate Chymical Remedies that are wont to be sold in Shops', could be obtained directly from the 'Spirit of Man's Blood'. This, he explained in *Some Considerations Touching the Usefulnesse of Experimental Naturall Philosophy* (1663), could be done by taking the blood 'of an healthy Young man as much as you please', and, whilst still warm, adding twice its weight in pure alcohol and sealing the mixture in a glass vessel for six weeks. Then, by heating it, both a salt and an oil would result, both being useful in the treatment of asthma, epilepsy, fever, pleurisy and tuberculosis.

A friend of Boyle's, 'an Excellent Chymist', apparently performed 'rare Cures' with this very remedy. Nevertheless, Boyle noted the difficulty of getting hold of decent quantities of blood from a healthy man, as well as what he considered the 'groundless abhorrency' some people had for such a medicine. For this reason he experimented with blood from deer and sheep, which he suggested promised to be almost as effective.[97] Nothing, however, appears to have come of this. More popular were Goddard's Drops, good for 'Diseases of the Head, Brain and Nerves': they were made from an oil chemically extracted from human bones, for the secret recipe of which Charles II allegedly paid a handsome sum.[98]

As the researches of Boyle and his contemporaries show, the seventeenth-century interest in chemistry extended beyond the search for the philosopher's stone. But this, as we shall now see, remained a discovery of keen concern to Boyle, as well as to some of his colleagues at the Royal Society. It was not a peripheral activity; it was part of the very revolution in science.

§

The early Royal Society's membership was as varied as its intellectual interests were wide. It was the embodiment of Francis Bacon's vision of Salomon's House, and John Graunt memorably described it in 1662 as the King's 'Parliament of *Nature*'.[99] Such was the appeal of chemistry and the search for the philosopher's stone to many of the society's first

Fellows that in 1703 John Pickering – who claimed friendship with Thomas Herbert, Earl of Pembroke, a former president – suggested that this 'Royal Academy' had been assembled by Charles II, Robert Boyle 'and other Great and Ingenious Practitioners' to search for the 'great Medicine'. By this he no doubt meant the elixir or the Liquor Alkahest – though the search, Pickering noted, had been 'without success'.[100]

That Robert Boyle, Sir Kenelm Digby (1603–65), the mathematician and architect Sir Christopher Wren (1632–1723), and other early Fellows of the society had all previously been members of a philosophical club known as the 'invisible college' has led some historians to see the influence of Rosicrucianism at work here; but this, like the manifestoes themselves, is probably a claim without substance.[101] Nevertheless, the interest that many Fellows had in chemistry was real enough. The society's secretary, Henry Oldenburg, was a keen supporter of chemistry, considering it, if 'rightly used', to be 'the best possible key' for discovering 'the admirable Treasures of nature'.[102] He also appears to have shared the Paracelsians' hope of prolonging life with chemical medicines. Between 1657 and 1660 he made a long tour of Europe, and his correspondence is filled with references to continental chemists and physicians who were searching for Helmontian medicines.[103] A memorandum in the Royal Society's archives records that, in a now lost letter he sent to Robert Southwell in January 1660, Oldenburg set out his 'opinion of the universal medicine and of those who live very long'. This, he noted, depended upon 'God's decree', and required 'a naturally strong body without any lapse into drunkenness and venery'.[104] Judging from Southwell's positive reply, Oldenburg had set out a fairly convincing argument in favour of chemistry.[105]

In 1670, in his capacity as Secretary to the Royal Society, Oldenburg translated and published in the *Philosophical Transactions* a letter from Jean Pierre de Martel, Professor of Anatomy at the University of Aix-en-Provence, 'concerning a way for the Prolongation of Humane Life'.[106] Martel began his letter by conjecturing that, were we 'more intelligent' in our understanding of 'the causes of a meerly natural Death' (as opposed to death by disease), 'we might procure for our selves an Age of continual Youth'. Having cited the 'illustrious *Bacon*' and the theory of maintenance and repair posited in *The History of Life and Death*, Martel likened 'the Engine of our body' to 'a Chymists Furnace'. Whilst at first it well retained its own natural heat, and was 'very proper for the operations of Art; but at last, chinks and crevices being made therein, it ceases to be so'. Martel, however, offered no explicit suggestions for chemically prolonging life. He referred only to the way 'the life of many dying persons' could be 'maintain'd, for some time', by making them drink hot, spirituous liquors. Yet Martel's bald conclusion was that 'there is no reason to despair of finding out such Medicins' as would one day fulfil Bacon's dream of longevity.[107]

Fellows of the Royal Society pursuing similar interests in chemistry included Thomas Henshaw and Sir Robert Paston. In the 1660s they collaborated, in an attempt to produce a 'red elixir'. They failed, and in 1671 Henshaw felt that Paston had too great expectations: 'I have often told y[o]u before, I have no hopes from Chymistry but to obteine an Extraordinary Medicine which will cure most diseases and maintin a vigorous health to ye time appointed.'[108] Henshaw, who went on to become Vice-President of the Royal Society in 1677, felt that many of Paston's hopes were castles in the air. But an 'extraordinary medicine', a life enhancing elixir, seemingly was not.

Sir Kenelm Digby, who had discussed the prolongation of life with Descartes in Paris in the 1640s, was another Royal Society Fellow hopeful of discovering 'extraordinary medicines'. As a young man he had made a great impression at court by healing a nobleman's injured hand with a 'powder of sympathy'. Francis Bacon, he later claimed, had even intended to add this medicine 'by way of Appendix, to his Natural History'.[109] When Digby's beautiful young wife Venetia died suddenly in 1633, it was rumoured he had accidentally poisoned her by making her drink 'viper wine', a chemical decoction made from the flesh of adders.[110] This medicine, it was claimed in a contemporary handbook, 'strengthens the Brain, Sight, and Hearing, and preserveth from Gray-hairs, [and] reneweth Youth'.[111]

The distraught widower retreated to Gresham College, London, where – as John Aubrey records – he 'diverted himself with his chymistry'.[112] Following the lead of Paracelsus and Jacob Duchesne, the grieving Digby spent his time experimenting with attempts to revive dead plants and animals. Later he actually claimed to have succeeded in bringing crayfish back to life from their crushed ashes.[113]

Whatever we may now think of such chemical ambitions, Digby cannot be considered a backward-looking philosopher. His treatise *On Bodies* has been described as 'one of the first fully developed atomist systems of the seventeenth century', and in 1652 he was placed by Isaac Barrow of Trinity College, Cambridge, alongside Descartes, Bacon and the French experimental philosopher Pierre Gassendi as contemporaries who had 'renewed ancient thought and struck out new paths in natural philosophy'.[114]

Indeed, Descartes may also have regarded chemical medicines as a way of prolonging life. As we saw at the end of the previous chapter, Descartes appears to have been relying largely on diet and will power to prolong his life to five hundred years. But, in an ambiguous remark, Adrien Baillet, the philosopher's seventeenth-century biographer records that Descartes 'required great Caution in the administering [of] Chymical Remedies'. This suggests that he used them at least on occasion. Like the Paracelsians, Descartes was opposed to phlebotomy, and had studied chemistry as well as anatomy in the 1630s. In a perhaps telling analogy, he likened the heart to an alchemist's flask, heating the blood and forcing it around the body, where it gave off spirits that, he suggested, then affected the body/soul relationship.[115] The historian Richard B. Carter has pondered the possible influence of what he calls 'doctor–alchemists' such as Paracelsus and Bernardino Telesio on Descartes. He even suggests that 'there is an aspect of Descartes's thought that seems to have been in remarkable harmony with the work and doctrine of the alchemists'.[116]

Descartes's premature demise did not chasten Sir Kenelm Digby in his quest for life-prolonging medicines. In 1654 he was considering investing the enormous sum of £700 in a proposal by Samuel Hartlib's son-in-law, the chemist Frederick Clodius, to establish a 'universal laboratory' that would 'redound . . . to the health and wealth of all mankind'.[117] In 1660, the year the Royal Society was founded, Hartlib recorded that Digby 'hath been up and down in Germany [searching] for the liquor alkahest the great elixir &c'.[118] Plagued by gout and the stone, Digby died five years later, following a violent fever. He was only sixty-one years old, and he had failed to find the elixir. His remains were buried in London alongside those of his beloved wife.

§

Despite the potentially materialist implications of the corpuscular theory being developed by Boyle and his contemporaries, a religious perspective on mortality remained strong. Broadly speaking, the seventeenth-century project to prolong life was not considered impious – though there were some critics who believed it was. In 1638, the London physician Tobias Whitaker noted there was a certain (unnamed) sect 'that stiffely defend the fatality of mans life; and that no man can bee preserved, prolonged, or restored'. The basis for this (presumably Calvinist) belief, Whitaker explained, was that death is 'an inexpugnable necessity determined of God', and was thus 'immutably fixed'. Whitaker considered this view 'Dangerous and impious'. For if it was true, what would be the need of the Church, or of individuals, to pray for the restoration of life or health, 'and to what end or purpose was the gift of healing dispenced to the Phisician, if death and dissolution of every kind bee predestinated'?[119]

Sir Kenelm Digby appears to have kept his options open on this matter. If he could not ensure the immortality of his flesh by chemical means, he could at least fall back on the permanence of the soul. A devout Catholic, he published a treatise in which he sought to prove the immortality of the soul from ordinary experience. Only the life of the body, he argued, could be shown to end. The soul lived on, and death was not the end of the individual essence.[120] Digby's example is thus informative. Even the devout believer in life after death could aspire to an endless *terrestrial* existence.

This seemingly ambivalent position is buoyed up by ideas expressed in a letter to Robert Boyle from another chemist and prominent figure in the Hartlib circle, Benjamin Worsley (1617/18–77). In the late 1650s, Worsley told Boyle that any sure-grounded 'Reformation . . . of the Art of medicine must in some measure know what is the Roote of death in every man'. The root of death for Worsley was still sin. Overcome sin (through faith), Worsley suggested to Boyle, and you would overcome death. In Worsley's opinion, if 'all the Gates & Avenues of death' were 'rightly' known, 'wee should not thincke it either Enthusiasticke or Ridiculous either to affirme or to expect a freedome [or] Liberation from the common state of mortality & corruption'. But Worsley did not confine his argument to faith alone: as there were numerous plants and animals that kill men, 'soe the Lord hath put a power in other simples [i.e. plants] to strengthen & quicken it'. The 'generality of Phisitians', Worsley believed, had mistakenly 'sought out the medicinall properties of things in a blended & confused manner': another way of searching might prove more fruitfull. Worsley did not name this method, but, as he reiterated throughout his letter to Boyle, he was certain death was neither 'absolutely fatall' nor 'necessary'.[121]

Worsley's was probably a radical opinion. But detailed study of thousands of early modern probate records from south-east England has led historian Ian Mortimer to suggest that, whilst people continued to recognize the importance of God in sickness and healing, 'the importation of Paracelsian ideas and the increase in the use and availability of chemical medicines provided an alternative to spiritual physic. In this context, the probate accounts almost certainly demonstrate that, as the medical strategies became available, they were taken advantage of by the dying.'[122] Increasingly, the seventeenth-century sickbed was a medicalized environment. The doctor was as likely as the priest to be in attendance to the end, in the hope that the latest medicines might postpone the final reckoning.

The sums expended in this pursuit of more life could be immense: in his last three months alive (November 1649 to February 1650), the Earl of Pembroke ran up a bill with his apothecary equivalent, in modern terms, to around £22,000.[123] Both Charles II and his Archbishop of Canterbury, Gilbert Sheldon (1598–1677) took chemical medicines in their final illnesses, in the hope of staving off death. And Moses Stringer, who taught chemistry at Oxford in the 1690s, produced numerous chemical medicines in London, including a so-called '*elixir renovans*' inspired by Paracelsus's 'renovating quintessence', which was intended 'to renew youth very much and help Old Age'.[124] Like a number of seventeenth- and early eighteenth-century proprietary medicine manufacturers, he made a considerable profit, which suggests that there were numerous willing customers. By 1703, Robert Pitt (1653–1713), a Fellow of the Royal College of Physicians, complained that the medical professions had 'sunk into the Craft of deceiving', with the promotion of fashionable but ineffectual medicines. Whilst the poor were cheated 'of their Lives and their Mony', the rich thought themselves 'very Fortunate' that they were able to purchase exotic-sounding medicines with the power to control disease, 'and make their Lives, if they use it often, almost Immortal. They pity the Vulgar, who have Dy'd before them, being not able to pay the Ransom.'[125] The whole medical profession, it appeared, had descended into disrepute.

Though deeply religious, Boyle does not appear to have had a theological problem with the possibility of a vastly prolonged life on earth. As Michael Hunter explains, for Boyle 'alchemy appeared to offer an empirical bridge between the natural and the supernatural realms which might provide irrefutable evidence of God's existence'.[126] Inevitably, his researches failed. In 1670 – and only in his mid-forties – he suffered a severe stroke. Aided by servants and secretaries, he continued to experiment and publish. His fame was such that his house became a place for sightseeing intellectuals: a painted board over the front door explained politely the hours when he did and did not receive visitors.[127] Boyle died on the last day of 1691, following years of declining health. He was only a few weeks short of his sixty-fifth birthday.

§

To what extent did the seventeenth-century search to prolong life enter the public consciousness? When Dr John Smith (1630–79), a Fellow of the College of Physicians in London, gave the subject lengthy consideration in his *Pourtract of Old Age* in 1676, he felt that 'the retarding of Age, the prolonging of Life, [and] the renewing of Youth' were subjects which 'have scarce entred the thoughts of Vulgar Pretenders to Physick'.[128] Like Bacon, Descartes and Jean Pierre de Martel, Smith believed that one day 'such noble Medicines may be found out and prescribed, that may innovate the strength of all the parts of old men'. He even thought it possible that humans, like insects, might one day be able to shed their skins and metamorphose. But he warned that, whilst '[s]ome means' may yet be found by physicians 'for the proroguing' of the diseases of old age 'and keeping them off for a time; and for the mitigation of their violent assaults, but for the total preventing, or the absolute curing, let no man living hope for'.[129]

By the close of the century, belief in the prolongation of life – like the belief in transmutation and in a universal medicine – was waning. In the early 1690s, the German physician Engelbert Kaempfer travelled to Japan, where he served as doctor to the Dutch embassy. He recorded an audience with the emperor in Tokyo, at which he and his companions were asked what he considered 'a thousand ridiculous and impertinent questions'. The emperor enquired '[w]hether our

European physicians did not search after some Medicine to render people immortal, as the Chinese Physicians had done for many hundred years'. Kaempfer replied that 'very many European Physicians had long labour'd to find out some Medicine, which should have the virtue of prolonging humane life, and preserving people in health to a great age'. He told the emperor that the best was a 'Spiritous Liquor' recently discovered by the Dutch chemist Franciscus Sylvius, and that it could be had in Batavia. But Kaempfer appears to have been mocking the emperor: he wrote elsewhere in his book that Japanese attempts to find a universal medicine had met 'with as good Success, as our European Philosophers can boast to have had in their searches after the Philosophers' Stone'. That is, the best of them had lived no longer than their eighties.[130]

Around the same time Kaempfer was being quizzed by the Emperor of Japan, Richard Bentley (1662–1742), a young Fellow of Trinity College, Cambridge, addressed the question of physical immortality in a sermon given at Oxford. The sermon was financed by a bequest in Robert Boyle's will, established with the purpose of confuting atheism, deism and Islam. In the first of these regular 'Boyle Lectures', Bentley tackled the atheists' claim that, although the human body is 'pretty good in its Kind', it does not look like 'the Workmanship of so great a Master' as Christians claimed. If God was such a good Creator – the argument went – why are humans 'so subject to numerous Diseases, so obnoxious to violent Deaths; and at best, of such a short and transitory Life'? It appeared humans were no better 'than some ordinary piece of Clock-work with a very few motions and uses, and those continually out of order, and quickly at an end'.[131]

Bentley admitted that these 'complaints about the Distempers of the Body and the Shortness of Life' were forceful. But he had an answer: the atheists' criticism reflected only their joy in luxury and pleasure. 'No question if an Atheist had had the making of himself,' Bentley retorted, 'he would have framed a Constitution that could have kept pace with his insatiable Lust, been invincible by Gluttony and Intemperance, and have held out vigorous a thousand years in a perpetual Debauch.'[132]

It is *our* fault, Bentley explained, that our lives are short and painful, not God's. When Adam was first created, he had been endowed 'with all imaginable Perfections of the Animal Nature'. It was through disobedience and sin that 'Diseases and Death came first into the World'. Furthermore, most of the distempers from which we now suffer 'are of our own making, the effects of abused Plenty, and Luxury, and must not be charged upon our Maker'.[133] As for the atheists' claim that life is too short:

> Alas for him, what pity 'tis that he cannot wallow immortally in his sensual Pleasures! If his Life were many whole Ages longer than it is, he would still make the same Complaint . . . For Eternity, and that's the thing he trembles at, is every whit as long after a thousand years as after fifty. But Religion gives us a better prospect and makes us look beyond the gloomy Regions of Death with Comfort and Delight. . . . We are so far from repining at God, that he hath not extended the period of our Lives to the Longævity of the Antediluvians; that we give him thanks for contracting the Days of our Trial, and receiving us more maturely into those Everlasting Habitations above, that he hath prepared for us.[134]

Bentley's lectures were subsequently published as *A Confutation of Atheism* in 1692. This future Professor of Divinity had drawn all the Christian moral arguments together and hammered one nail into the coffin of seventeenth-century dreams of immortality. Rationalist science briskly knocked in two more.

The astronomer and mathematician Edmond Halley (1656–1742), a prominent Fellow of the Royal Society, struck one. Using records from the Silesian town of Breslau, in 1694 Halley made the first advancement upon John Graunt's famous 1662 study of London's Bills of Mortality. Acknowledging a debt of inspiration to Bacon's *History of Life and Death*, Graunt had calculated that 36 per cent of those born in the capital died before they were six years old, whilst only around 7 per cent 'die of *Age*' – that is, over seventy years old.[135] Halley now showed that in Breslau the average life expectancy at birth in the period 1687 to 1691 was a mere 33.5 years. This rose to 41.55 at five years old, but then declined slowly after that. At eighty, one's life expectancy was a little less than six more years. No one lived beyond a hundred.

Though flawed in their conception, Halley's calculations were the first statistical attempt to compute the average span of human life – and he was dismissive of any vision of its vast prolongation. He pointed out that these data showed 'how unjustly we repine at the shortness of our Lives, and think ourselves wronged if we attain not Old Age; whereas it appears hereby, that the one half of those that are born are dead in Seventeen years time'. Instead of 'murmuring at what we call an untimely Death', Halley advised that 'we ought with Patience and unconcern to submit to that Dissolution which is the necessary Condition of our perishable Materials, and of our nice and frail Structure and Composition: And to account it a Blessing that we have survived, perhaps by many Years, that Period of Life, whereat the one half of the Race of Mankind does not arrive'.[136]

Thomas Burnet (*c*.1635–1715) struck another scientific nail into the casket of seventeenth-century prolongevism. A former Fellow of Christ's College, Cambridge, and Master of Charterhouse school, in his controversial *Sacred Theory of the Earth* (1684), Burnet argued that, before the biblical Flood, the earth had been smooth and featureless, like an egg. At the deluge, this pristine globe had cracked open to release the underground floodwaters, and it had shifted off its axis: what had once been a world of perpetual springtime had become one of contrasting seasons. Since then, the earth had been a ruinous, mountainous wreck of its former self. Burnet thus added ingenious weight to the argument already advanced by Bacon and others, that on this decayed, less fertile earth men could no longer live as long as they once had. His reason why Methusaleh and the patriarchs had lived so long thus echoed the one advanced by Bacon: they had 'liv'd longer' because

> [t]hat food that will nourish the parts and keep us in health, is also capable to keep us in long life, if there be no impediments otherwise; for to continue health is to continue life; as that fewel that is fit to raise and nourish a flame, will preserve it as long as you please, if you add fresh fewel, and no external causes hinder . . .

Added to this was what he called the great stamina of the patriarchs. This, and their nourishing diet, had been irreversibly lost when the Flood changed the world out of all recognition.

Burnet thus dismissed what he called 'those Projectors of Immortality, or undertakers to make Men live to the Age of *Methusalah,* if they will use their methods and medicines'. He explained that there was 'but one method' by which that could be achieved, and that was

> [t]o put the Sun into his old course, or the Earth into its first posture; there is no other secret to prolong life; Our Bodies will sympathize with the general course of Nature, nothing can guard us from it, no Elixir, no Specifick, no Philosopher's-stone. But there are Enthusiasts in Philosophy, as well as in Religion; Men that go by no principles, but their own conceit and fancy, and by a Light within, which shines very uncertainly, and, for the most part, leads them out of the way of truth.[137]

Burnet did not believe anything could be done to overcome this catastrophe, and he had no time for anyone who suggested that they could.

✪

CHAPTER 3

The Romantic Error[1]

The thought of making men immortal here is vain, as every thing must be which tends to counteract the course of nature, and the purpose of God. We blush to think men could propose, and men receive the doctrine: and yet philosophers and chymists have pretended to it. 'Tis not too harsh to say the first were fools; the latter cheats; or both enthusiasts too wild for truth or reason.

John Hill, *On the Virtues of Sage, in Lengthening Human Life* (1763)

Paris, Autumn, 1793 The Reign of Terror – and the city is awash with fear, excitement, celebration, blood. Tricolours and coloured placards hang at the windows, declaring 'Unity, liberty, equality, fraternity or death'. Streets vie to outdo one another in displaying their patriotism: it has become dangerous, says one Parisian, 'to be considered less revolutionary than your neighbour'.[2]

The year had started with the execution of the king. On 7 October the Revolutionary Calendar is adopted: Year II is deemed to have begun on 22 September, first anniversary of the abolition of the monarchy. On 16 October the king's widow, Marie Antoinette, follows him to the scaffold: hair shorn, hands tied, she is carried in a cart through the Paris streets to the Place de la Révolution. Bound to a plank, her neck clamped, she is swung beneath the guillotine and its heavy, rattling blade falls. It is intended to be a swift, efficient, humane despatching. The suggestion that the head might actually *survive* decapitation –

albeit only for some few seconds – will not come for a couple of years yet.[3]

Why all this bloodshed? At war with itself and its neighbours, the French Republic is fighting to survive: 'liberty must prevail at any price', declares one of the confident young Jacobin leaders, Louis de St-Just. 'You must punish not merely traitors but the indifferent as well.' The ship of the Revolution will reach safe harbour, he explains, 'only by ploughing its way boldly through a Red Sea of blood'.[4] The English journalist William Hazlitt recalls later how the whole of France 'seemed one vast conflagration of revolt and vengeance . . . The shrieks of death were blended with the yell of assassin and the laughter of buffoons.' Whole families were led to the scaffold for no other offence than their relationship to those convicted of crimes against the state.[5]

Those whose heads roll from the guillotine include aristocrats, failed generals, fallen politicians, revolutionaries and counter-revolutionaries, royalists, deserters, captured soldiers, food hoarders, traitors, the wealthy, the poor, the guilty and the innocent – men, women, priests, bishops, nuns. At the direction of the Committee of Public Safety, nearly 3,000 are executed in Paris. In the provinces, 14,000 more are killed – beheaded, drowned, shot to death by canon. 'What a sight!' writes an onlooker after a massacre of royalists: 'Worthy indeed of Liberty!'[6] On 31 October, twenty-one members of the moderate revolutionary party, the Girondins, are executed in Paris. They go to their deaths singing the *Marseillaise* and shouting '*Vive la République!* The crowd applauds as each head falls.[7]

In hiding somewhere in the city, fearful of the same fate, is one of the Girondins' most famous, most respected supporters. Marie Jean Antoine Nicolas Caritat, Marquis de Condorcet, was born in 1743 to a noble family and educated – like his hero, that 'ingenious and bold philosopher', René Descartes – by Jesuits. Condorcet is one of France's most gifted mathematicians: elected at twenty-six to the Académie des sciences, in 1785 he became its permanent secretary. Another of his heroes is that 'philosophical genius', Sir Francis Bacon, and in 1775 he writes proposals for the advancement of scientific learning along lines 'similar, as it were, to Bacon's *Atlantis*'.[8] Science would be put to work, as Bacon had dreamt, for the public good. In Condorcet's scheme, a universal scholarly language will link an international republic of scientists.

Publishing a wide array of treatises on mathematics, philosophy and education, he becomes a member of the Académie française and is elected a Fellow of the Royal Society of London. By the 1780s, his Parisian home is a salon for intellectuals and radicals. 'Converse with him,' writes one female admirer, 'read what he has written; talk to him of philosophy, belles-lettres, science, the arts, government, jurisprudence, and when you have heard him, you will tell yourself a hundred times a day that this is the most astonishing man you have ever heard; he is ignorant of nothing . . . in fact, nothing is beneath his attention and his memory is so prodigious that he has never forgotten anything.'[9] Condorcet is against slavery, monopolies, war and the death penalty; he stands for free trade, parity in wealth and equality in gender, and education and happiness for all.

When the Revolution comes, he welcomes it: he is elected to the Commune of Paris, then to the Legislative Assembly, eventually

Figure 3.1 Marie Jean Antoine Nicolas Caritat, Marquis de Condorcet: photograph by E. Desmaisons after a lithograph. Described by one contemporary as 'the most astonishing man you have ever heard', the French revolutionary and philosopher spent the last months of his life penning his *Sketch for a Historical Picture of the Progress of the Human Mind* (1795). The final chapter envisioned a future free from poverty, inequality and disease – and perhaps even death.

becoming its president; he joins a sub-committee of the Committee of Public Safety. Opposed to the execution of the king, when his educational schemes are rejected, he turns his attention to devising a permanent constitution for the Republic. This too is rejected, and in a public letter he attacks the Jacobin Constitution proposed by Robespierre, St-Just and their powerful followers. Indignant, the Jacobins order his arrest for treason. In July, Condorcet goes into hiding. As the Terror takes hold, he writes the work for which he will be best remembered: the *Sketch for an Historical Picture of the Progress of the Human Mind.*

The *Sketch* was written in urgency, on the backs of old proclamations and other pieces of scrap paper, and is riddled with repetitions and mistakes. Condorcet divides human history into nine epochs. These chart the progress of society, from its origins in superstitious barbarism on through the progress of science, learning, reason and enlightenment to the buoyant dawn of the tenth era – its rising sun, marked by the founding of the French Republic. The Terror is but a momentary regression – for, as a contemporary will note of the *Sketch*, even here Condorcet 'speaks about the Revolution with nothing but enthusiasm'. He never abandons faith in the movement that has condemned him, considering this episode 'only as one of those personal mishaps nearly inevitable in the midst of a great movement productive of general happiness'.[10]

It is Condorcet's vision of the tenth era, '[t]he future progress of the human mind', that is so remarkable, and so relevant to the story of this book. For Condorcet's belief is 'that nature has set no terms to the perfection of human faculties; that the perfectibility of man is truly indefinite; and that the progress of this perfectibility . . . has no other limit than the duration of the globe upon which nature has cast us'.[11] Condorcet, like the other great French *philosophes* who helped to found the Age of Reason, is filled with optimism that the world can be refashioned. It is an ambition not without just cause. At the same time as the guillotine is falling in the Place de la Révolution, government schemes are afoot for providing free medical assistance to the elderly and unfit; an education act is passed on 6 January 1794 that will provide three years of free, compulsory primary education to the young citizens of France; on 4 February slavery is abolished in the French dominions. As St-Just declares, in the new Republic all will be equal:

'we must have neither rich nor poor'.[12] 'Happiness', he explains, 'is a new idea in Europe.' Inspired by the revolutionaries of the recently founded United States of America, it is a project whose time has finally come.[13]

And, as Condorcet now writes in hiding, 'all the means that ensure' the perfection of the human race 'must by their very nature exercise a perpetual influence and always increase their sphere of action'. From this single fact of constant, inevitable progress, Condorcet predicts the eventual 'perfectibility of man':

> No one can doubt that, as preventative medicine improves and food and housing become healthier, as a way of life is established that develops our physical powers by exercise without ruining them by excess, as the two most virulent causes of deterioration, [extreme poverty] and excessive wealth, are eliminated, the average length of human life will be increased and a better health and a stronger physical constitution will be ensured. The improvement of medical practice, which will become more efficacious with the progress of reason and of the social order, will mean the end of infectious and hereditary diseases and illnesses brought on by climate, food, or working conditions. It is reasonable to hope that all other diseases may likewise disappear as their distant causes are discovered. Would it be absurd then to suppose that this perfection of the human species might be capable of indefinite progress; that the day will come when death will be due only to extraordinary accidents or to the decay of the vital forces, and that ultimately the average span between birth and decay will have no assignable value? Certainly man will not become immortal, but will not the interval between the first breath that he draws and the time when in the natural course of events, without disease or accident, he expires, increase indefinitely?[14]

The prolongation of life and the possibilities it might engender had seemed a fascinating possibility to earlier philosophers of the French Enlightenment. 'I don't ask for immortality,' Denis Diderot had told Jean d'Alembert, fellow editor of the mighty *Encyclopédie*, 'but just give man twice the length of his present life, and you can never tell what might happen!'[15] But Diderot had died in 1784, aged seventy, and d'Alembert a year earlier, at sixty-six. Their younger colleague

Condorcet, however, *is* suggesting that man might one day be immortal, because the end of the indefinite progression he predicts cannot be charted. As it gradually increases down the centuries, there is no time, he writes, by which this process of improvement 'must of necessity stop'. So, he explains, 'we are bound to believe that the average length of human life will for ever increase'. The only impediment to its illimitable progress is 'physical' (by which Condorcet presumably meant environmental rather than political) 'revolutions; we do not know what the limit is which it can never exceed. We cannot tell even whether the general laws of nature have determined such a limit or not.'[16] This is not the restitution of some lost Golden Age: no longer are philosophers looking backwards, as they had in Bacon's day, to Eden. They are looking forwards, to a future unlike anything that had ever been imagined before.

With his friends the Girondins languishing in prison, these optimistic thoughts, Condorcet writes, are 'an asylum, in which the memory of his persecutors cannot pursue him', and where he can forget 'man tormented and corrupted by greed, fear or envy'.[17] He completes his essay, as he notes on the manuscript, on 'Friday 4 October 1793 old style, 13th of 1st month of the Year Two of the French Republic'.[18]

Only the day before he had been tried *in absentia* by the revolutionary tribunal and sentenced to death.[19] For almost six months more he remains in hiding. Then, by a decree of 13 March, he is outlawed. A new wave of the Terror has begun: Robespierre is annihilating his own enemies as well as those of the Republic. Fearing for his safety and that of those hiding him, Condorcet dons disguise and flees Paris. He seeks refuge with old friends at Fontenay des Roses: frightened, they forsake him. On 27 March he is recognized as an aristocrat in a country inn: according to legend, he gives himself away by ordering an omelette to be made with a dozen eggs. Only an enemy of the people could make such a brazen request! Imprisoned at the Bourg-Egalité, he is found dead on the morning of 29 March. Aged only fifty, either exhaustion or self-inflicted poison had killed him.[20]

Three more months of bloodshed, and the Jacobins finally fall. Robespierre, St-Just and ninety of their supporters are sentenced to death. 'You monster spewed out of hell', one woman shouts as Robespierre is led to the Place de la Révolution. 'The thought of your execution makes me drunk with joy.'[21] With Robespierre guillotined, the Convention votes for funds to print and distribute Condorcet's *Sketch* throughout France. It is published in 1795, and one contemporary immediately hails it as the 'apocalypse of the new gospel'.[22] Condorcet's vision of indefinite progress towards eventual physical immortality on earth is a glimpse of the secular state's new Jerusalem.

§

It had appeared at the close of the seventeenth century that hopes for physical human immortality were over. What had changed to make Condorcet so optimistic? The eighteenth century had opened with the same scepticism with which the old one had closed: there was no way the human being was meaningfully going to prolong his or her life. Certainly, the respected natural philosopher and clergyman William Derham (1657–1735) accepted the great ages of, among others, Methuselah and Thomas Parr in his oft-published Boyle Lectures, *Physico-Theology, or, A Demonstration of the Being and Attributes of God from his Works of Creation*.[23] But old age was still seen, and experienced, as a melancholy period of deterioration and decay. As Lady Sarah Cooper wrote in her diary in 1713, approaching her seventieth birthday: 'My facultys are broken by the infirmitys of age, flat and dull, irksome and tedious, apt to nothing but complaint under the weight of one evil or another that befalls mee. The powers of soul and body are in a languishing condition.'[24] In her mid-seventies, Sarah Churchill, Duchess of Marlborough, reflected in 1735: 'At my age I cannot expect to continue long, nor have I anything left to make me desirous of it.'[25]

The most famous articulation of such eighteenth-century geriatric pessimism occurred in a novel. Its author, Jonathan Swift (1667–1745), Dean of St Patrick's Cathedral, Dublin, was one of the foremost wits of the early Georgian era. In his youth Swift had been secretary to Sir William Temple, the man who, in his 'Essay upon health and long life', had observed that the natives of Brazil were said to have lived two or three hundred years, and who had himself met a 124-year-old beggar in Staffordshire. Temple believed that disease, and thus the need for doctors and strong medicines, was due largely to our indolent lifestyles and imprudent diets. His advice for a long and healthy life included baths and massages, as well as medicines – not from apothecaries or physicians, but from natural herbs and remedies: garlic, quinine, powdered millipedes in butter, tobacco. The two greatest blessings of life, Temple believed, were health and good humour, each depending upon the other. Whether long life was a blessing, only God could determine: 'But So much I doubt is certain; that in Life as in Wine, He that will drink it good, must not draw it to the Dregs.'[26]

Swift saw things in much the same way. For, rather than becoming healthier and longer, the lives of men and women – in England at least – appeared to be getting sicklier and *shorter*. From a peak of 42.7 years in 1581, life expectancy at birth in England was actually falling. It reached a nadir of just 25.3 in 1726 – the year in which *Travels into Several Remote Nations of the World, by Lemuel Gulliver*, was first published.[27]

In his wicked satire, Swift's luckless sailor finds himself a visitor at the Grand Academy of Lagado: this is nothing less than Swift's parody of the ambitious new science of Francis Bacon and the experimenting natural philosophers of the Royal Society. After visiting Laputa, Gulliver travels on to Luggnagg, where he is asked whether he has met any '*Struldbruggs* or *Immortals*'.[28] It is explained to him that these immortal men and women are occasionally born by chance, and can be identified by a red circular spot on their foreheads. Gulliver cannot hide his excitement at hearing such incredible news:

I cryed out as in a Rapture; Happy Nation where every Child hath at least a Chance for being immortal! Happy People who enjoy so many living Examples of antient Virtue, and have Masters ready to instruct them in the Wisdom of all former Ages! But, happiest beyond all Comparison are those excellent *Struldbruggs*, who being born exempt from that universal Calamity of human Nature, have their Minds free and disengaged, without the Weight and Depression of Spirits caused by the continual Apprehension of Death.[29]

He is surprised not to have met any of these immortals at the king's court, considering that they would make 'such wise and able Counsellors'. If it had been *his* luck to have been born a Struldbrugg, Gulliver reflects, he would have spent two centuries acquiring wealth, and would then have become a recorder of events and of the changing customs of the times: 'By all which Acquirements, I should be a living

Treasury of Knowledge and Wisdom, and certainly become the Oracle of the Nation.' He would create a 'lodge' of a dozen of his 'own immortal Brotherhood', and together they would serve their nation. They would

> remark the several Gradations by which Corruption steals into the World, and oppose it in every Step, by giving perpetual Warning and Instruction to Mankind; which, added to the strong Influence of our own Example, would probably prevent that continual Degeneracy of human Nature, so justly complained of in all Ages.[30]

He would be endlessly entertained by the rise and fall of nations, by discoveries in science such as the universal medicine, 'and many other great Inventions brought to the utmost Perfection'.[31]

But, as Gulliver quickly discovered, dream and reality were much at odds. Struldbruggs soon wearied of immortality: by the age of eighty 'they had not only all the Follies and Infirmities of other old Men, but many more which arose from the dreadful Prospect of never dying'. They were 'opinionative, peevish, covetous, morose, vain, talkative', and as repositories of history they were hopeless, having 'no Remembrance of any thing but what they learned and observed in their Youth and middle Age'. Anyone seeking 'the Truth of Particulars of any Fact' was 'safer to depend on common Traditions than upon their best Recollections'.[32]

Growing old had no benefits for Swift. As in the ancient Greek myth of Eos the Titan, who asked Zeus that her human lover Tithonus be allowed to live forever, but forgot to ask that he remain forever youthful, immortality was a curse not to be wished for. Better, perhaps, to ignore it altogether – like Louis XV, who forbade all mention of death in his presence.[33] Swift's own old age was a sorry tale of physical and mental deterioration, his 'understanding' in his final years 'quite gone'.[34] He died, distracted, some two years short of his eightieth birthday.

§

By the early eighteenth century, alchemy and chemistry had evolved into two distinct subjects: one was hailed as serious science, the other, increasingly ridiculed as preposterous folly. In 1744, in his *Guide to Health Through the Various Stages of Life*, Dr Bernard Lynch dismissed all the claims of the alchemists. He accepted that the patriarchs had been long-lived and that there had been exceptional modern cases of longevity. He noted the long lives of Parr and Jenkins – as well as a more recent case: a woman named Margaret Paten from Paisley in Scotland, who had died at St Margaret's workhouse, Westminster, on 26 June 1739, aged 138. But these were anomalies; Lynch assured his readers that the natural human lifespan since the Flood was a mere eighty years.[35]

Ageing, Lynch explained, was a natural result of the body's everyday actions and functions. Studies of nerves and veins under the microscope showed this to be the case: the small vessels are slowly destroyed, the humours stagnate and grow thick, fibres of the body adhere together. The subtle internal juices waste and dissipate, digestion weakens, nourishment becomes more and more deficient, and the humours that carry the nourishment for life slow and fail. Thus all the 'great Promises' that had been made

> with Assurance, as to the Prolongation of Life for so many hundreds of Years, are vain and imaginary, being unsupported both by Reason and Experience: Of this Tribe are *Van Helmont's Primum Ens, Paracelsus's Elixir Proprietatis*, the *Primum Ens* of Animals; all those precious *Liquors*, that *potable Gold*, those Conserves of *Rubies, Emeralds, Elixirs of Life*; that fabulous Fountain, that was reported to make People grow young, cannot hinder us from Decay and Old Age. Nor is it likely that Life should be prolong'd even by the best Methods in Nature, so many Years as the Chymists pretend by their Art; but their own Experience is a Proof of their Temerity and Inability herein.[36]

James Mackenzie (1682?–1761), honorary Fellow of the Royal College of Physicians of Edinburgh, praised Lynch's book, considering it both 'full and perspicuous'. Yet, in his *History of Health, and the Art of Preserving it* (1758), Mackenzie also commended Sir Francis Bacon and the Dutch chemist Hermaan Boerhaave (1668–1738) for recommending 'new and bold methods to prolong life'. Mackenzie noted that a 'sure and easy road to longevity', different from the general rules of health based on the six non-naturals, 'seems to be among the desiderata in our art, the discovery of which is reserved, perhaps, for a more meritorious generation'. Like Lynch, he held out no hand for those who hoped science and medicine might one day prolong life. In his opinion, 'the greatest efforts of the human mind to extend a vigorous longevity much beyond fourscore, will generally prove ineffectual'.[37]

Yet there were still a few men prepared to speculate upon chemistry, and others ready to believe them. The most famous of all, who claimed to have lived for hundreds of years, perhaps made his earliest recorded appearance to Robert Boyle. In 1679 a Parisian doctor of medicine by the name of 'Mr de St Germain' or 'Mr Sangermain' visited London, where he called on Boyle. Boyle asked St Germain to find out more about a man named Georges Pierre des Clozets, with whom Boyle had been in recent correspondence. Returning to France, St Germain wrote back that he had done as Boyle had asked, and confirmed that Clozets was indeed 'intimate with the most illustrious figures of this century', that he 'has the honour of being the disciple of certain among them', that 'he spends his time in the pursuit of fine things, and that he is not lacking in [transmuting] powder'.[38] Clozets clearly was, as he claimed, a well connected alchemist.

By 1685, Boyle and St Germain had met again. Back in Paris, and with the help of instructions from an unidentified but 'expert friend' of Boyle, St Germain was attempting to prepare the philosopher's stone. But his experiments failed, and St Germain told Boyle he was disappointed his friend had not given him 'all the necessary details' for completing the process. He asked Boyle to obtain further information 'so that you may make certain regarding all the facts which we now have in our hands'.[39] This is the last we hear of St Germain in Boyle's correspondence. Six years later Boyle was dead.

The name of St Germain does not reappear until 1745. In that year the author Horace Walpole wrote to a friend that the authorities in London had recently arrested 'an odd man who goes by the name of Comte St Germain. He has been here these two years,' Walpole explained, 'and will not tell who he is or whence, but professes . . . that he does not go by his right name. He sings and plays the violin wonderfully, is mad, and not very sensible.'[40]

If this was the same St Germain who had met Boyle, he would have had to have been in his nineties. Unless he truly *had* perfected the elixir, he cannot have been the same man – for this same St Germain turned up again, in France, twelve years later. There he demonstrated a range of chemical skills – including the manufacture of dyes and synthetic gems. The French king, Louis XV, was impressed, and provided St Germain with a laboratory in which to work. And it was at this time that rumours of St Germain's great age started circulating. Karl Heinrich Baron von Gleichen, who met St Germain in Paris in 1759, collected detailed evidence:

> While not all the fables and anecdotes relating to the age of Saint-Germain merit the attention of serious people, it is true that the collection I have made of testimonies of persons of good faith, who have attested the long duration and almost incredible preservation of his person, has in it something of the marvellous. I have heard [the composer Jean-Phillippe] Rameau and an elderly relative of a French Ambassador to Venice affirm having known Saint-Germain there in 1710, looking like a man of fifty, and Monsieur Morin, who has since then been my secretary at the Embassy, for whose veracity I can speak, told me that he knew Saint-Germain in Holland in 1735, and was prodigiously astonished at finding him, now, not aged by so much as a year . . .[41]

These reliable witnesses appeared to confirm that, at the very least, St Germain had not aged at all in almost half a century!

Gleichen observed that, though St Germain possessed various 'chemical secrets' for making colours and dyes, he never heard him speak of a universal medicine. He noted, though, that the *comte* 'kept a very strict regime': he would never drink whilst eating, and purged himself with a preparation of senna, a natural laxative. Disappointingly, 'that was all he had to recommend to those who asked him what they should do to prolong their lives'.[42]

By 1760 St Germain was living in the Low Countries. In that year, Count Kriegsrath Kauderback of Saxony, the King of Poland's resident representative at The Hague, recounted rumours he had heard of St Germain's incredible age and youthfulness. 'What is certain', he told a friend, was that a member of the States General, who was then approaching seventy, had told him that as a child he had seen 'this extraordinary man in the house of his father . . . and yet that he has the agile, loose movements of a man of thirty'. He had a full head of hair and 'hardly a line on his face', and it was noted that St Germain ate only chicken, cereals, vegetables and fish.[43]

Stories of the *comte*'s great age and youthfulness continued to circulate. In 1777, Count Ernst Heinrich Lehndorff, Chamberlain to the Queen of Prussia, recorded that St Germain denied suggestions he had lived since the time of Christ, stating that it was others who invented such 'myths'. But, Lehndorff acknowledged, St Germain 'admits to having lived a long time' and that he was 'not thinking to die yet'. St Germain told Lehndorff that 'anyone following his way of life should attain considerable age, without illness'. Lehndorff added that the *comte* 'follows a very strict diet, studies great frugality, drinks only water, never wine, and takes only one light meal a day'. Lehndorff also noted that the *comte* took a diuretic in his tea that tasted of aniseed, and that '[h]e discourses constantly of [the] right balance between body and soul. When this is observed, he says, the life-machine cannot get out of order.'[44] St Germain, it appeared, confirmed the ancient theory regarding diet and the six non-naturals – rules most people observed only in the breach. He was, perhaps, like Thomas Parr – an exceptional example of what could be achieved by true moderation.

Exactly how old St Germain actually claimed to be is unclear. According to Count Maximillian von Lamberg, St Germain told him he was 350 years old. Von Lamberg also stated that the *comte* carried an autograph book containing personal messages from historical figures such as the sixteenth-century French essayist Montaigne.[45] Giovanni Casanova, the famous Venetian philanderer and raconteur, recorded a number of meetings with the *comte*, the first at a dinner party. Instead of eating, St Germain had spent the entire meal talking. According to Casanova, he had claimed to possess a water that he gave to ladies, 'not to make them younger, for that, he said, was impossible'. Rather, it would 'preserve' them and keep them from ageing any further. At the same meal, St Germain 'declared with impunity, with a casual air, that he was three hundred years old, that he possessed the universal medicine', and that he could create diamonds. Casanova considered this 'very singular man' to be 'the most barefaced of all impostors', but admitted: 'I found him astonishing in spite of myself, for he amazed me.'[46]

Such was the *comte*'s reputation that in 1777 Count Philip Karl von Alvensleben, Prussian Ambassador to Saxony, was asked by Frederick the Great to make further investigations. Interviewing St Germain – who, he thought, looked about seventy years old – Alvensleben asked probing questions. In one answer, the *comte* alluded to Swift's book, *Gulliver's Travels*. As Alvensleben reported to Frederick: 'I said that if the secrets he claimed could bring me to believe what people said of his great age', what he had just said 'of the pleasure with which he had read Swift, made me doubt it, unless indeed he knew ancient times down to the last particulars'. St Germain replied that he knew these details 'from the very circumstantial accounts' written by those who had lived in that time.[47] These remarks are rather vague. Was the *comte* confessing that he was playing a game, as of course he was – and admitting that he knew no more about the past than what anyone can learn simply by reading history books?

The Comte de St Germain remains a tantalizing enigma. He straddles two epochs of science and rationality, being somehow credible, yet not altogether so. He illustrates the depth of the human wish to believe that very long life *is* a possibility. But on 27 February 1784, like all men, St Germain died. Impoverished, he was buried a few days later at Nikolaikirche, in Eckernförde, Denmark, where he had spent the last four years of his mysterious life. The exact cause of death, like the date of his birth, remains unknown.[48] It would probably have been classified under that early modern catch-all: 'old age'.

§

The emphasis St Germain placed on diet reflects a growing trend in the eighteenth century. Late in the seventeenth century London had become the largest city in Europe, and it continued to grow. Its expansion marked the emergence of England as the richest nation in the world, the first to experience both agricultural and industrial revolutions. Wealth and health became inextricably linked. As the London physician William Stukeley put it in 1722, 'Our leaving the country for cities and great towns, coffeehouses and domestick track of business, our sedate life and excesses together' were preparing 'a plentiful harvest' of 'nervous' disorders, ranging from gout to melancholy and hypochondriasis.[49]

In 1733, the Scottish physician George Cheyne (1671/2–1743), a Fellow of the Royal Society and of the Edinburgh College of

Physicians, memorably named this 'the English malady'. England was getting richer, but the 'luxury' and 'sensibility' accompanying that wealth – fine clothes, racy fashions, too much work or too much idleness, and exotic foods and drugs such as tea, chocolate, coffee, tobacco, sugar, spices and opiates – were all combining to make the upper and middling classes morally and physically weaker. It was what doctors had been complaining of a century and more earlier, only worse. Much worse.

Medicine was seemingly unable to keep up. Whilst the 'scientific revolution' of the seventeenth century had seen spectacular advances in physics, chemistry, mathematics and – in particular – astronomy, medicine had lagged way behind. By 1704, Dr Richard Mead, one of the wealthiest and most renowned of London's physicians, was fretting that his profession 'still deals so much in conjecture, that it hardly deserves the name of a science'.[50] In 1725, Sir Richard Blackmore, also a Fellow of the Royal College of Physicians, thought it lamentable 'that this useful and important Art should be improved so little in so many Centuries; and that its State should still continue so uncertain and imperfect. We have hitherto discovered *few Remedies* of a peculiar specifical Virtue, for the Cure of *any* Diseases.'[51] The only great development would be Edward Jenner's discovery of vaccination for smallpox in the 1790s. Smallpox was a major killer, and, once it struck, there was little an early modern doctor could do. Such diseases had to be left to run their course. Under these circumstances, it is little wonder that few people looked to medicines for the prolongation of life, and prevention was considered better than cure.

Something, however, could be done to fight nervous disorders, and Dr Cheyne sought a way to solve this rising epidemic of misery and melancholy, of ill-health and premature death: he blamed diet. And, like Luigi Conaro and St Germain, he saw strict regimen as the route to extending our lives in comfort and happiness.

Cheyne had an intimate relationship with food and drink. Studious by nature, sedentary by habit, he hailed from what he called a family 'disposed . . . to *Corpulence*'. In 1702 he moved from Edinburgh to London; in the sociable eating and drinking culture of the English capital's abundant clubs and coffeehouses, he quickly (as he later recalled) 'grew excessively *fat, short-breath'd, lethargic,* and *listless*'. Ballooning in weight to a startling 32 stone (448 pounds), in 1706 he suffered a nervous breakdown.[52] But by following a careful diet of vegetables and water Cheyne shed considerable weight: he would tell his friend, the novelist Samuel Richardson, that around this time he shed 16 or 18 stone 'of my rotten Flesh'.[53]

A leaner, fitter Cheyne established himself in the fashionable spa resort of Bath. There he treated the wealthy, fashionable sick and wrote his *Essay on Health and Long Life*, published in 1724. A contemporary soon observed that the book 'is now almost in every body's hand': within a year it had reached a seventh edition, and it was still in print a century later.[54] In his *Essay*, Cheyne explored each of the six 'non-naturals' of Galenic medicine, explaining how air, diet, sleep, exercise, evacuations and the passions of the mind all played their part in ensuring a long, healthy life. Luigi Cornaro, 'Old' Parr and Henry Jenkins made appearances as examples of just what could be achieved by abstemious diet and regular exercise.

Cheyne explained that the '*Groundwork*' for his method had to be 'laid, carried on, and finished in *Abstemiousness* . . . All the rest will be insufficient without this. And *this* alone, without *these*, will suffice to carry on Life, as long as by its natural Frame it was made to last.'[55] Those who wantonly and habitually transgressed the self-evident 'Rules of Health' were guilty of nothing less than 'Self-Murder'. Gout, melancholy and a painful, premature death were the price paid for those who ignored them.[56] Medicine played only a small role in Cheyne's practice. Though he sometimes prescribed chemical remedies, he believed that the most 'pernicious Error in *Physic*' had been the one, 'first introduc'd and propagated by *enthusiastical Chymists*, *Quacks* and *Symptom-Doctors*', which suggested that life-enhancing medicines could be created in the laboratory. Such chemical medicines might help cure topical diseases, but they could not aid general health or longevity. With all the guilt of a reformed fat man, Cheyne stated that food was everything: 'Diet alone, proper and specific Diet, in Quantity, Quality and Order', he declared, was 'the sole *universal Remedy*, and the only Mean known to Art . . . which can give *Health*, *long Life* and Serenity.'[57] As he told Samuel Richardson, 'You had as good shoot yourself as alter your Diet.'[58]

Ideally, this diet would be vegetarian, and he recommended gruel, milk porridge and rice pudding. Whilst a student at Oxford, John Wesley noted that 'the famous Dr Cheyne' advised drinking two pints of water a day, and one of wine.[59] But Cheyne believed that those who drank *only* water from their youth 'would live probably till towards an hundred Years of Age, that being the Term of Life appointed by the Design of the Creator' after the Flood.[60] As he wrote in his *Essay on Regimen* (1740), water was the 'true and universal *Panacea*, and the *Philosophers' Stone*'.[61] It was not much of a diet. As one witty critic quipped: 'For my Physician I accept your Book; But, by the *Gods!* – you shall ne'er be my *Cook.*'[62]

Of course, Cheyne was hardly the first advocate of abstemiousness or vegetarianism. Cornaro had recommended the former, whilst Thomas Tryon had vigorously promoted the latter. Yet Cheyne was one of the most successful eighteenth-century promoters of a meat-free diet. He did not, however, expect all his clients to follow his advice: white meat was acceptable, even if seeds, fruits and salads were better.

Given the extremeness of Cheyne's views, it is perhaps surprising that his books were so popular. Yet his followers included both the aristocracy and the literati. The valetudinarian poet Alexander Pope tried following a meagre Cheynean diet supplemented by mineral water, and recommended the Scottish physician to friends. But it appears he could not always stick to the regime. As the writer George Lyttelton wrote approvingly – if jokingly – to Pope in 1736: 'Immortal Doctor Cheyney bids me to tell you that he shall live at least Two Centuries by being a Real and practical Philosopher, while such Gluttonous Pretenders to Philosophy as You, Dr Swift and My Lord Bolingbroke die of Eating and Drinking at Fourscore.'[63]

Another literary advocate of Cheyne's dietetic method was Samuel Johnson. 'Do not read this slightly', he advised his friend Hester Thrale in 1775, when recommending Cheyne's *Natural Method of Curing the Diseases of the Body, and the Disorders of the Mind Depending on the Body*: 'you may prolong a very useful life'.[64] Thrale clearly enjoyed reading Cheyne. As she noted in her diary in August 1790:

Few Books carry so irresistible a Power of Perswasion with them as Cheyne's do; when I read Cheyne I feel disposed to retire to *Arruchar* in the Highlands of Scotland – live on Oat bread & Milk, and bathe in the Frith [sic] of Clyde for seven Years; and I do partly believe that was I to take up that impracticable Resolution at the Turn of Life . . . I should last a healthy Woman to a Hundred years

old – Absence of all Passions, Fish now & then, but a continual Diet
of good Seeds & Milk, would with a little Bark for chewing, with
Rhubarb if Occasion arose, give one amazing Strength:– compared
to the Artificial Fire excited by Meat & Wine.[65]

Thrale attributed the origin of her own good health to Cheynean
principles: 'My early Habits of extreme Temperance have doubtless
lent me my present Vigour', she mused to herself: 'early Abstinence did
surely lay the Foundation of a robust Constitution rarely equalled'.[66]
After Thrale remarried she retired to a rural idyll in Wales, but died at
the age of eighty in 1821, following a serious fall.

There were, inevitably, critics of Cheyne's method. One anonymous
doctor – like Cheyne, a Fellow of the Royal Society – rushed into print
with a reply to the *Essay on Health and Long Life*. This author reckoned
that there must already have been above two hundred books written on
the general subject of health and long life. (Cornaro's sixteenth-century
tract, for example, was enjoying considerable popularity, going
through some fifty editions in English in the eighteenth and nineteenth
centuries.)[67] But, whilst such works might please some, they could
'never save them from Death': even before Cheyne, senescence and
death had 'baffled Ages of Physicians'.[68] Cheyne had nothing particularly new to offer, this unknown Fellow argued. And besides, he
pointed out, death was necessary to keep the population of the earth's
'crowded Nations' in balance. If Cheyne's vegetable diet really 'had
Power to make it, that none but People of a Hundred and Twenty
should die . . . Nature would sink under her own Burden, and . . . Men
would then die of Scarcity, as they now do of Plenty'. But these were
merely quibbles: the Fellow claimed that his own book would show
that Cheyne's 'Trick for making Men Immortal upon *Asparagus* and
Parsnips, will not deserve a Patent'.[69] Diet alone could not so prolong
life: there was still a role for the physician.

By the late 1730s, Cheyne's ideas were becoming more extreme.[70]
His *Essay on Regimen*, published in 1740, clearly displayed his
advancing religious mysticism: human life, he explained, was nothing
but God's punishment for sinfulness. In five 'discourses' on medical,
moral and philosophical matters, he explained that the world was
'really and literally' nothing more than a place 'of Banishment, of Pains
and Punishment'. The earth is to the human race as Siberia is to the

Russians, the Bastille is to the French, or Newgate is to the English: we are all 'Prisoners, Slaves, and Felons' upon a 'defaced and spoilt Planet'. Like labourers at the oars of a slave galley, our sinful life is 'a malignant Fever'. And things got no better with time, Cheyne explained: old age, 'the *Cardinal* Disease of human Life . . . finishes the dark Scene of human Misery with perpetual Aches, Sores and Infirmities of Body and Mind'.[71]

Yet if God is infinitely perfect, how could He have created anything so 'imperfect or unhappy'? It was a puzzling question. But 'it is Fact', Cheyne observed, 'that all sentient and intelligent Beings' on earth 'are universally more or less, miserable and imperfect'. Given that all these physical ills clearly come from eating meat, why did we do it? Since God had given permission to eat meat only after the Fall, He had clearly intended it 'as a Curse, or Punishment'. By eating meat, God had made humans '*feel* and experience the natural and necessary Effects of their own Lusts . . . by painful and cruel Distempers'. His purpose had been to 'shorten the Duration of [our] natural Lives, that *Sin, Misery*, and *Rebellion*, might not increase infinitely'.[72] It was a mechanical explanation for a theological action. It was clear to Cheyne 'that this Indulgence for animal Food' did not occur until 'the *Aera* of *Longævity*' – of Adam, Methusaleh and Noah – had ended. This was at the time of the Flood, when the tyranny of mankind had forced God 'to begin a new World'.[73]

While Cheyne considered his *Essay on Regimen* to be 'the best, most useful, and solid Work I ever composed',[74] now, almost seventy years old, he was rather losing direction. Though impressed (and surprised) to have lived as long as he had, given his own early excesses, his weight had once more ballooned. In a letter to Samuel Richardson written in December 1741, Cheyne complained that, once again weighing some 450 pounds, he was 'overgrown beyond any one I believe in Europe'.[75] 'The great fat Doctor of Bath', as one poet called him, died a wealthy man less than eighteen months later.[76]

§

Up to the early 1700s, English-speaking theorists appear to have dominated the study of the prolongation of life, and accounts of great longevity were more common in the British Isles than elsewhere in the

West. In the late seventeenth century, the French Ambassador to The Hague told Sir William Temple 'that in his Life he had never heard of any Man in *France* that arriv'd at a hundred Years'.[77] Temple was surprised by this, and found it difficult to explain. He supposed it had something to do with the fact that the French lived life too *well* to live it long. A short and merry existence was, he reflected, 'without doubt better than a long with Sorrow or Pain'. The Swiss anatomist and botanist Albrecht von Haller, who was familiar with the story of Henry Jenkins, wrote in the mid-eighteenth century that 'England seems to exceed all other nations in the number of those who live to an advanced age'.[78]

As if to mark the growing continental fascination with longevity, in 1757 the distinguished author Bernard le Bovier de Fontenelle, one-time secretary of the Academy of Letters, died almost exactly one hundred years after his birth. A French centenarian, at last! Appropriately, the mid-century saw another satire of the debate on the prolongation of life sitting alongside Swift's: 'there is no hope left of immortality in this body, or even of prolonging our lives to three or five hundred years', wrote the Westphalian doctor Johann Heinrich Cohausen in *Hermippus Redivivus* (1742). Cohausen (1665–1750) claimed to have discovered an account of one Clodius Hermippus, who in classical times had lived 115 years and five days by inhaling the breath of young women. This feat, Cohausen suggested, merited consideration from both physicians and posterity.[79] *Hermippus Redivivus* went through numerous editions and translations between 1742 and 1847: that it was taken seriously by some, who failed to recognize it for the satire it was, shows the level which the interest in the prolongation of life had reached.

It has been suggested that the French Enlightenment saw a growing respect for the aged in France, reaching a 'sentimental extreme' by the time of the Revolution.[80] Certainly more names were found to be added to the lists of French longevity, for the eighteenth century was the great age of collecting and cataloguing, anything from insects to antiquities. In 1761 the Parisian publisher Augustin-Martin Lottin (1726–93) compiled an *Almanac of Old Age*, listing individuals who had lived for a hundred years or more.[81] Even a French equivalent of Thomas Parr was found: Annibal Camous of Nice, who was said to have been 121 years old when he died in 1759, supposedly having prolonged his life with herbal applications.[82] A belief in the possibility of long life – even if such a life was uncommon – clearly penetrated popular consciousness. Thus in 1787 a teenage Napoleon Bonaparte wrote to the Swiss physician Samuel Auguste Tissot, seeking a cure for his seventy-year-old uncle's gout: 'people live to a hundred years and more', young Bonaparte explained, 'and my uncle, by his constitution, ought to be of the small number of those privileged'.[83] Forty years later Bonaparte would remark that, considering the progress that had been made in science, a way would eventually be 'found to prolong life indefinitely'.[84]

This interest in the prolongation of life was gathering pace elsewhere in Europe. The apparently increasing evidence concerning long lives justified the impression that most humans died far too early. In Denmark, the civil servant Bolle Willum Luxdorph (1716–88) spent much of his spare time gathering information on the very aged. He bought books containing anecdotes and collected portraits of men and women who had lived beyond eighty years; he corresponded with Scandinavian clergymen, asking them to search their parish registers for information on centenarians. The product of these labours, published in 1783, was probably the first fully illustrated directory of human longevity. It contained 254 examples of men and women aged between 80 and over 180.[85]

Of course, Luxdorph's directory included those great English centenarians Parr, Jenkins and Bayles. But there were numerous continental examples, too. They included an Hungarian peasant, Petracz Czartan, who had died in 1724, supposedly at the age of 185. Czartan's claim rested on his presence at an historically verifiable event: as a boy he had (he said) witnessed the Ottoman Turks' capture of the Hungarian fortress at Temesvár in 1552. Luxdorph also recorded the case of Christian Jacobsen Drakenberg, who had been making claims of his great age since as early as 1732. In that year he had obtained a certificate from Cornelius Nicolae, vicar of the church in Skee (now a part of Sweden). This testified that, according to the church register, Drakenberg had been born on 18 November 1626 and baptized there, which 'proved' he was already 106 years old. Five years later, a Danish nobleman paid for the old man's wedding, and for the remainder of his long life Drakenberg received an annuity from the Danish king. To have passed himself off as being aged 106 in 1732, Drakenberg must already have been fairly old. Yet he lived for another forty years – dying in 1772, supposedly aged 145 years, ten months and two days.[86] His story, like 'Old' Parr's, soon spread around Europe. And, like Parr's, it was faithfully accepted by many, recurring in numerous subsequent accounts.

A catalogue similar to Luxdorph's was published in England in 1799. Its author was James Easton, a shopkeeper (possibly a bookseller) in Salisbury. He produced a whole directory containing the names, ages, places of residence, years of decease and brief biographies of numerous men and women who, over the past 1,733 years, had lived for a century or more. Easton noted that he had 'most scrupulously refused admittance' of any account on the authenticity of which he 'had the smallest doubt'. Thus he explained that he had omitted a claim made by a Portuguese author about an Indian who had died in 1566 aged 370. His list included Cornaro, Jenkins and Fontenelle. Parr was there, his 'premature' death suggesting to Easton that a modern human lifespan of 200 years was not impossible. The death of Petracz Czartan at the age of 185 seemed to confirm this: 'The man, who has reached the farthest extent of mortal existence', Easton confidently proposed, 'may be considered as a pattern of human nature, in its utmost perfection, and as an instance of what is possible to be attained under favourable circumstances.'

In Easton's opinion, it was 'next to impossible, that he who leads a regular and sober life should fall sick, or die a natural death before the time that Nature has prescribed; for distempers cannot be produced without a cause; and if there be no bad one reigning, there can be no fatal effect, or violent death'.[87] It was an optimistic view, but, in an age before the germ theory of disease, it made some sort of sense. However, as David Troyansky has shown, for most of those who – either by luck or constitution – lived long in the eighteenth century, old age was a struggle, in which physical decline most often resulted in hardship and destitution.[88]

§

A more rigorous mind than Easton or Luxdorph's was that of the French mathematician and natural historian George Louis Leclerc, Comte de Buffon. Born in 1707 to a wealthy family and, from 1739, Keeper of the Jardin du Roi in Paris, Buffon was a man with an extraordinary array of interests. In true Enlightenment fashion he published a huge, multi-volumed history of the natural world – where, amongst many other things, he speculated on the great age of the earth, which he considered much older than the 6,000 or so years of Christian tradition. In his essay 'Of old age and death', which was part of his book *On Man*, Buffon reflected on the likely length of human life-spans. If only to keep himself out of trouble with the Catholic Church, he accepted the remarkable longevity of the patriarchs. This achievement, he explained, had no doubt been due to the youthfulness of the earth: then it had been less solid and compact (since gravity had only acted upon it for a short time), and it was also moister. This meant that the patriarchs' muscles and bones were softer, suppler, and they only developed fully after 120 or 130 years; the first humans had thus lived proportionately longer. For Buffon, the stages of human growth and development were now fixed, in the same way that the maximum length of life was. Death was simply the last stage in a long process of decline: in the final stages, being dead was not much different from being alive – it was all just a question of degree. It was for this reason that some people awoke from death after burial.[89]

Buffon's most innovative and enduring legacy, however, was his attempt to calculate with precision the natural length of human life. He believed that this could be determined according to how long a species took to reach adulthood. A dog, for example, reaches maturity in two years, and lives in total for about ten. A man, by contrast, takes sixteen to eighteen years to reach maturity. So, using the same integer of five that determines the maximum life of a dog, Buffon reckoned that a human could expect to live a 'natural' span of between eighty and ninety years.[90]

Senescence and death were, for Buffon, inevitable; there was nothing to be done to alter the rigid laws of nature. The ideas of visionaries about the prolongation of life, he declared, should have died with them. The universal medicine, blood transfusion and all other means of rejuvenating and rendering humans immortal, were chimerical. If the body was well constituted, it might be possible to add a few years to life. Moderation in the passions, temperance and sobriety might help, but even this was doubtful.[91] If we consider all the different races and classes of men, he wrote, we find that 'the difference of race, of climate, of food, of comforts, makes no difference in the duration of life, it will be seen at once that [it] depends neither upon habits, nor custom, nor the quality of food, that nothing can change the fixed laws which regulate the number of our years'.[92]

Yet Buffon also admitted that, compared to other animals, humans live for a long time, and it was undeniable that their lifespans were less certain and more variable. In the conclusion to his discussion of long life in *On Man*, Buffon produced tables and examples of long life, pondering on how many more years a human of a certain age could expect to live. Though he felt that the 'mechanical law of nature' meant that our food or other habits would have little effect on our maximum lifespan, he observed that there were examples of high-altitude dwellers living longer – for example in the mountainous regions of Scotland, Wales, Switzerland and the Auvergne.[93] He appears even to have accepted the account of Drakenberg's 145 years. He referred to it in a supplement to his *Natural History* in 1777, noting also the examples of a Turin man named Andre-Brisio de Bra, who had lived to be 122, the Sieur de Lahaye, who had died at 120, and of Istwan Horwaths, a former cavalry officer, who had died in Lorraine in 1775 aged 112.[94] Satisfyingly, however, Buffon lent appropriate weight to his own theory that the 'natural' human lifespan was only eight or nine decades. He died in 1788, only seven months after his eightieth birthday.

§

Buffon proved to be one of the most influential Enlightenment thinkers on natural history. His readers included Albrecht von Haller (1708–77), Professor of Anatomy, Surgery and Medicine at the University of Göttingen. In *First Lines of Physiology* (1751), Haller expanded a little on Buffon's attempt at a scientific analysis of ageing, explaining death as an increasing process of rigidity and 'earthiness' resulting in 'gouty concretions', 'brittleness of the bones' and bony, stony 'crusts' which coat the arteries. This accumulation of damage culminated in the 'rigidity of the whole body, the decrease of the muscular powers, and the diminution of the senses'. Together, these 'constitute old age; which sooner, or later, oppresses mortals severely'. Again, there were qualifying factors. Old age came sooner if the body was 'subjected to violent labour, or addicted to pleasures, or fed upon an unwholesome diet; but more slowly, if they have lived quietly and temperately, or if they have removed from a cold to a warm climate'.

Towards the conclusion of this ageing process, Haller explained, 'the necessity of natural death approaches'. He believed, however, that death from natural exhaustion of the body happened 'but rarely'. Rather, diseases or accidents terminated most lives prematurely, and very few people reached a natural end. 'One in a thousand exceeds the age of 90', he reckoned, whilst 'one or two perhaps in a century live to the age of 150'. Haller was uncertain of the cause of these exceptional survivals, though he speculated that a temperate climate might explain the numerous cases of English longevity. He added that sobriety, a temperate diet and a 'peaceable disposition' with 'a mind not endowed with very great vivacity, but cheerful, and little subject to care', might all be helpful.[95]

But Haller was not saying all that much that was original. More interesting was the work being undertaken in an exciting new scientific field: electricity. In the 1750s, some of the best research was being carried out by the American printer, inventor and revolutionary diplomat Benjamin Franklin (1706–90). Franklin was also interested in the prolongation of life: as a young man, he had toyed with Thomas Tryon's vegetable diet, and he continued to be fascinated by the idea that humans might one day live for prodigious lengths of time. As he told his friend, the English chemist Joseph Priestley, in 1780:

> The rapid progress *true* science now makes, occasions my regretting sometimes that I was born so soon. It is impossible to imagine the height to which may be carried, in a thousand years, the power of man over matter . . . all diseases may by sure means be prevented or cured, not excepting even that of old age, and our lives lengthened at pleasure even beyond the antediluvian standard.[96]

Priestley, too, had pondered over such a possibility. In his *Essay on the First Principles of Government* (1768), he argued that political liberty had raised modern people far above primitive societies. Likewise, mankind in future ages would surely be superior. Government, society, science and liberty would all play instrumental parts in cementing this progress, for *knowledge*, as Lord *Bacon* observes, being *power*, the human powers will, in fact, be increased; nature, including both its materials, and its laws, will be more at our command; men will make their situation in this world abundantly more easy and comfortable; they will probably prolong their existence in it, and will grow daily more happy . . . Thus, whatever was the beginning of this world, the end will be glorious and paradisiacal, beyond what our imaginations can now conceive.[97]

Things would not prove so happy for Priestley, however. At the height of the French Revolution, a Birmingham mob picked him out as a republican sympathizer. They destroyed his laboratory and scientific papers, and Priestley fled, eventually ending up in the United States. He died there in 1804, at the respectable age of seventy.

Franklin remained fascinated by the possibility that progressive science would some day solve the problem of death. In 1773 he told another of his friends, the French *philosophe* Dr Jacques Barbeu du Bourg (1709–79), how he had discovered three flies apparently drowned in a bottle of Madeira wine that had been shipped from Virginia to England. By exposing them – as Boyle had once done – to the warmth of the sun, two had revived. Reflecting on this curious feat of preservation and revival, Franklin told his Parisian friend:

I wish it were possible . . . to invent a method of embalming drowned persons, in such a manner that they may be recalled to life at any period, however distant; for having a very ardent desire to see and observe the state of America a hundred years hence, I should prefer to any ordinary death, the being immersed in a cask of Madeira wine, with a few friends, till that time, to be then recalled to life by the solar warmth of my dear country!

Aware, however, that they probably lived 'in an age too early and too near the infancy of science' to hope to see such an art 'brought in our time to its perfection', Franklin joked that he was content at this stage to see 'the resurrection of a fowl or a turkey'.[98] Of more possible relevance were Barbeu du Bourg's experiments with electricity. Franklin was particularly interested in what he described as the Frenchman's 'observations on the causes of death' and in the experiments he proposed 'for recalling to life those who appear to be killed by lightning'. But he noted 'that the doctrines of life and death in general are yet but little understood'.[99]

Of less appeal was the Austrian physician Franz Anton Mesmer (1734–1815), who caused a great stir with his claims to be able to cure disease and prolong life through his supposed control of 'animal magnetism'. Eventually settling in Paris, Mesmer attracted hundreds of patients – particularly women – from the upper echelons of society. In 1784, a commission was appointed to investigate his healing claims, with Franklin, who was then the US Ambassador to Paris, heading the committee. After five months of investigation, they debunked Mesmer's premises, stating that they were nothing more than the potentially dangerous action of the imagination.[100]

In November 1789, Franklin famously joked that nothing was certain in life 'except death and taxes'. He died a few months later, of pleurisy, aged eighty-four. He lived not quite long enough to hear of the electrical experiments by the Italian anatomist Luigi Galvani (1737–98). In 1792, Galvani described how he made the legs of dead frogs move by connecting them to a weak electric current. Eleven years later, in London, Giovanni Aldini applied galvanic electricity to the corpse of an executed murderer. It would be reported afterwards that 'the jaw began to quiver' and 'the adjoining muscles were horribly contorted, and the left eye actually opened'. Contractions to some of the muscles were so strong 'as almost to give an appearance of re-animation'.[101] Had the originating force of life finally been found? 'The phænomena of electricity', John Abernethy pointed out to the College of Surgeons in 1814, and that of life, 'correspond'. From the chemical experiments of Sir Humphry Davy, it appeared to Abernethy that electricity 'performs all the chemical operations in living bodies'. He did not mean to affirm that electricity was life, but the two certainly appeared to be closely linked.[102]

§

In 1797, the German doctor Christoph Wilhelm Hufeland (1762–1836) tried to draw some of these new ideas into a more rational approach to the prolongation of human life. Like Cheyne, Hufeland's patients included the literati: the most famous were Friedrich Schiller and Johann Goethe. In the preface to *The Art of Prolonging the Life of Man*, Hufeland wrote that long life 'has at all times been the chief wish, the principal object of mankind'. But, he complained, 'how confused and contradictory are all the plans ever proposed for obtaining it!' Hufeland proposed to set the study on a properly scientific, intellectual basis, overcoming what he called the objections of theologians, the limited vision of physicians and philosophers, and the false theories and vain promises of chemists and quacks such as the Comte de St Germain.[103]

For Hufeland, the life of man, when 'physically considered', was 'a peculiar chemico-animal operation'. Therefore, 'like every other physical operation', it must have 'defined laws, boundaries, and duration'. And, like every other physical operation, these could be 'promoted or impeded, accelerated or retarded'. Hufeland explained that his book, written in his leisure hours over the previous eight years, would 'bring this science back to solid and simple principles'. His principal aim was 'to establish the Art of prolonging Life on systematic grounds, and to make known the means for accomplishing that object'. This would be achieved by what he called 'dietetic rules and a medical mode of treatment for preserving life'. This was to be a new science, he claimed, and he called it 'macrobiotics': it was, he emphasized, distinct from the medical art of *promoting health*, and was to be specifically concerned with the art of *prolonging life*. It was, effectively, a return to Sir Francis Bacon's vision of the early 1600s, but based on more modern principals.

This debt to Bacon was no coincidence. Hufeland acknowledged the *History of Life and Death* as the model for his own study. Of all the previous theoreticians on the subject, from ancient times to the present day, it was the 'great Bacon, whose genius embraced every branch of science', for whom Hufeland had the greatest respect. Bacon's theories on this subject, in particular his renovating, refreshing and nourishing processes, were 'bold and new . . . The truth contained in these ideas cannot be denied; and, with some modification, these precepts might at all times be employed.'[104] Indeed, though Hufeland's *Art of Prolonging Life* would be the most thoroughgoing examination of the subject since the publication of Bacon's study almost two centuries earlier, his principles for the prolongation of life would not be so very different from the Jacobean Englishman's.

Thus, like Bacon, Hufeland began by attempting to understand and explain the 'vital power' or 'grand cause' of life. Only if we could comprehend 'the internal nature of that sacred flame' might we 'discover by what it can be nourished, and by what it is weakened', and thus answer the question of how to prolong it.[105] Yet Hufeland, like all those before him, had difficulty solving what he called this 'sanctum sanctorum of Nature'. The vital power that gives life to things was, he mused, a force somewhat like electricity, light, fire or magnetism – but more penetrating, more subtle and invisible. This vital power gave living things their different forms, and protected them from putrefaction; yet, even after death, this force persisted in another form. As soon as dead matter starts decomposing, he observed, 'its fine particles begin to be again animated in a thousand small worms, or to display their revival under the figure of beautiful grass'. This process was 'the great circle of organic life': the apparent death of live matter is 'only a transition to a new life; and the vital power leaves a body only that it might unite itself again with it in a more perfect manner'.[106] This was an immortal, materialist vision, in which life continued in an endless chain of being.

Hufeland identified four principal agents invigorating this 'vital power': they were heat, oxygen, light and – to a lesser degree – food.[107] At first, this vital power is fresh and energetic, and we grow and are able to regenerate. But gradually consumption becomes equal to renovation, and then the body's vital power decreases. This gradually results in physical decay, degradation and, eventually, total dissolution, concluding in death. Every created thing thus passes through three distinct periods: growth, stasis and decline. In Hufeland's theory – which, in fact, was not so different from that of Aristotle or Bacon – all methods and means for the prolongation of life could be brought under four basic principles:

1 The quantity of vital power in the original being.
2 The robustness and health of vital organs.
3 The rate at which vital power is consumed: for example, a candle lit at both ends will burn twice as fast, whilst a lamp burning oxygen will burn out ten times faster than a similar one burning 'in common air'. Thus, whilst exertion uses up vital power, rest replenishes it. Hufeland thus contrasted 'fast living' or 'intensive

life' with 'slow living' or 'extensive life': the less intensive the life, the longer its duration.
4 The efficiency of the means of regeneration.[108]

Having established these four factors, Hufeland proceeded to answer the great unresolved question: 'What is the absolute duration of human life?'

The most long-lived moderns, he suggested, were those who laboured in the open air, such as farmers, soldiers and sailors. Experience, Hufeland observed, 'incontestably tells us' that in these circumstance a life of 150 or even 160 years was quite possible. Recent examples included a Cornishman who had died in 1757 in his 144th year; a labourer in Holstein who had died in 1792 in his 103rd year; and an old soldier in Prussia who had died as recently as 1792, in his 112th year. But Hufeland's star examples were those seventeenth-century Englishmen, Thomas Parr (152) and Henry Jenkins (169). Indeed, for Hufeland, as for so many others, Harvey's autopsy of Parr 'proves that, even at this age, the state of the internal organs may be so perfect and sound that one might certainly live some time longer'. Hufeland concluded that it was possible 'with the greatest probability' to 'assert, that the organization and vital power of man are able to support a duration and activity of 200 years'.[109]

For Hufeland, this fact corroborated the biblical account of the long lives of the patriarchs. He did not take these literally. The patriarchal year, he explained, had been just *three* months long, not twelve. Previous authors had founded their dreams for human longevity on an error. Methuselah's 969 years should be quartered: he had in fact lived some 240 years – 'an age', as Hufeland observed, 'which is not impossible, and to which some men in modern times have nearly approached'.[110]

Nonetheless, the German physician observed that few people reached such extraordinary longevity as a quarter of a millennium. In the present age, he reckoned, not one in a thousand people lived even a hundred years. Hufeland drew the same conclusion as Haller had forty years before: most deaths occurring before the hundredth year are brought on artificially – 'by disease or accident'. His 'practical art' of prolonging life therefore depended upon identifying 'the friends and the enemies of life'. One who kept good company with the former would live long; 'but he who prefers its enemies, will shorten his existence'. The key to Hufeland's art of prolonging life beyond a hundred years thus consisted in distinguishing 'friends' from 'enemies', and in 'learning to guard against the latter'.

The 'enemies' of long life were numerous and formed a catalogue of all that might be considered wrong with civilized society. They included delicate nursing; physical excess in youth; over-exertion of the mental faculties; diseases and injudicious methods of treating them; suicide; impure air and city living; intemperance in eating and drinking; spirituous liquors; heightened passions (amongst which he included 'the dangerous stresses of business'); fear of death; idleness, inactivity, and languor; an overstrained power of imagination, the danger of 'imaginary diseases' and excessive sensibility; poisons and infections; and, finally, old age itself was an enemy of life. (The old, it might be noted, do not tend to live long. This was Bacon's trick for prolonging life: defer ageing and stay young for as long as possible. Prolonging old age is pointless.)

With his many enemies to long life, Hufeland reflects the lingering pessimistic side of the Enlightenment. As the influential French philosopher Jean-Jacques Rousseau had suggested, the only hope mankind had to improve its health was to escape civilization altogether: 'the numberless Pains and Anxieties . . . which the Mind of Man is constantly a Prey to', he had written in his famous *Discourse Upon the Origin and Foundation of the Inequality Among Mankind* (1761), 'are the Fatal Proofs that most of our Ills are of our own making, and that we might have avoided them all by adhering to the simple, uniform and solitary Way of Life prescribed to us by Nature'. The evidence of the happy and healthy 'savages' discovered around the world by European travellers was, to Rousseau, confirmation of his argument. Man in a state of nature was proof that 'this Condition is the real Youth of the World, and that all ulterior Improvements, have been so many Steps, in Appearance towards the Perfection of Individuals, but in Fact towards the Decrepitness of the Species'.[111]

Hufeland's 'friends of life' were more numerous than his enemies. Top of the list was good physical descent: this was important, and we will return to it at length in the next chapter. If your parents lived long, it was believed that you were likelier to live long yourself: Hufeland pointed out that Parr's great-grandson had recently died in Cork, at the age of 103.[112] But what if your parents had died young? There were other 'friends' of long life that could be implemented – though they were not much different from Bacon's. They included prudent physical education; an active and laborious youth; abstinence from sexual intercourse too young, and a deferred and happy marriage; plenty of sound sleep and lots of bodily exercise; enjoyment of fresh air in a 'moderate' climate; rural rather than city life; travel; cleanliness and care of the skin; proper food, with moderation in eating and drinking, and good dental care; mental tranquillity; prevention and judicious treatment of disease. Finally, if all these were not enough, there was the cultivation of mental and bodily powers.

These last were not to be underestimated. They formed the foundation of one of the last major Enlightenment reflections on the possibility of future human immortality – as we shall now see.

§

The belief that the human mind plays a key role in keeping us healthy and prolonging our lives was not new to the Enlightenment. Paracelsus had acknowledged the importance of 'joy', whilst in 1633 William Vaughan wrote that he considered 'mirth' to be the 'the principall naturall meanes' for prolonging life.[113] Luigi Cornaro had written of his great care never to abandon himself to melancholy, 'banishing' from his mind 'whatever might occasion it: I made use of all the Powers of my Reason', he wrote, 'to restrain the Force of those Passions, whose Violence does often break the Constitution of the strongest Bodies'.[114] This was part of his secret for a long life.

Robert Burton dissected the *Anatomy of Melancholy* at great length in 1621, whilst in the *History of Life and Death* Bacon paid close attention to 'the *Affections,* and *Passions* of the *Minde*', with the purpose of seeing 'which of them are Hurtfull to long Life' and which 'profitable'. He felt that hope (if it was not thwarted), leisure and contemplation were all beneficial, and noted that the ancient philosophers Socrates, Plato, Democritus and Seneca were all long-lived. He believed, however, that extreme emotions such as 'great Joy' shortened life, since they attenuated the spirits; great fear and suppressed anger were also bad – but worst of all was envy.[115] Edward Maynwaringe concurred with Bacon. In his *Preservation of Health, and*

Prolongation of Life, he eloquently summed up early modern thinking on the role of the mind in creating physical complaints. As he wrote of revenge, jealousy and envy in 1670:

> These Diseases of the *mind* are as painful Ulcers, continually lancinating, corroding or inflaming: they gnaw and eat like a *Cancer*, they take away the nourishment from *food*, and refreshment from *sleep*: the anguish of these *sores*, render every thing unpleasant and unserviceable for the wellfare and support of the Body: so that these *sicknesses* of the *mind*, make the *Body* to pine and languish, introducing a secret Consumption, wasting the Spirits and . . . enfeebling all the faculties.[116]

Hope, joy and mirth, by contrast, were to be embraced and cherished 'as the *supports* of your *life*'. For Maynwaringe, these 'enlivening affections' were 'the greatest friends to, & preservatives of *health* and *strength*'.[117] The long-lived man would be firm of heart and happy of mind. René Descartes also placed the power of the mind over the body at the centre of his scheme to prolong his own life vastly – and saw it as a cure for at least some physical ailments.[118]

Such reflections were not merely philosophical. They were grounded in anatomical investigation. By the mid-seventeenth century, the brain and nervous system were subjects of increasingly diligent enquiry. Descartes studied anatomy and, according to Richard B. Carter, believed that the living human was 'a compound organism that continues to live because the mind to which it is connected as a whole is concerned with health'. Carter suggests that through his 'inquiry into the defining borders shared by soul and body, Descartes revealed his grounds for unbounded confidence in the ability of every human being to correct himself . . . by employing certain characteristics of the brain and nerves'. This 'quasi-medical' inquiry, Carter argues, gave later theorists such as Rousseau 'what they considered objective, non-theological scientific foundations for their doctrines of the innate perfectibility of men'. It was 'Cartesian medicine that fed their optimism'.[119] In Oxford, meanwhile, the anatomist Dr Thomas Willis (1621–73) set about a thorough study of the brain, his aim being 'to unlock the secret places of Mans Mind'.[120] The work of the Cambridge mathematician and astronomer Isaac Newton, although he was not an anatomist, would also prove highly influential in redirecting Cartesian ideas on the mind–body relationship. His theories of gravity and of the role of action at a distance, together with his ground-breaking work on optics, proved spurs for further research, and prominent physicians, including the young George Cheyne, would apply Newtonian mechanical theories to the realms of life and disease.

The influence of Willis and Newton can be seen in the physician–philosopher John Locke's influential *Essay Concerning Human Understanding* (1690) and *Some Thoughts Concerning Education* (1693). In these works he presented the human mind as 90 per cent the product of nurture: 'nature', Locke argued, played only a small role in creating our personalities. Children were to be thought of as blank sheets of paper or pieces of wax, 'to be moulded and fashioned as one pleases'.[121]

According to another physician–philosopher, David Hartley, author of *Observations on Man* (1749), the progress of every individual 'in his Passage through an eternal Life is from imperfect to perfect, particular to general, less to greater, finite to infinite, and from the Creature to the Creator'.[122] Ultimate, unending perfection, for Hartley, would come only after death. But even in this life the mind could be *improved* – all to the benefit of the individual and society: the outcome, as the French Enlightenment *philosophes* argued, would be progress, happiness and the worldly perfection of both species and society.[123] On seventeenth-century foundations, a new science was being constructed in the eighteenth: psychology, defined by Harvey as 'the Theory of the Human Mind'.[124]

The most prominent thinker to develop these theories of the mind was the English philosopher and novelist William Godwin (1756–1836). Influenced by radical thinkers such as Rousseau, Locke and Hartley, and by the 'liberal and enlightened examples of human genius' which Swift had displayed in *Gulliver's Travels*,[125] at the age of twenty-seven Godwin abandoned his rural nonconformist ministry and moved to London, to write. In the second volume of his *Enquiry Concerning Political Justice, and its Influence on General Virtue and Happiness* (1793), he turned his consideration to what he called 'the sublime conjecture of Franklin, that "mind will one day become omnipotent over matter"'.[126] Godwin wondered: if this could be true of *all* matter, then why not over the matter of our own bodies? If over matter at ever so great a distance, why not over matter which, however ignorant we may be of the tie that connects it with the thinking principle, we always carry about with us, and which is in all cases the medium of communication between that principle and the external universe? In a word, why may not man be one day immortal?[127]

Certainly gravity could control planetary matter at the most extreme distances – and in 1704 the Newtonian physician Dr Richard Mead had even published a medical treatise on the supposed influence of the sun and moon on human bodies. But how, exactly, could *thought* control matter?

Figure 3.2 William Godwin: oil painting by James Northcote, 1802. 'We're all born to die, both the weak and the strong', wrote the poet John Courtney in 1794, 'Unless our existence sage Godwin prolong; He'll teach us by reason, Death's portals to batter.' Godwin's suggestion that our minds might one day compel our bodies to keep going forever struck a chord with contemporaries.

As an obvious example, Godwin pointed out that the mental intention necessary physically to move one's own hand has its origins in the brain – in thought. Likewise, involuntary emotional responses (such as fear, or lust) can display themselves as physical effects – such as the heart palpitations we might experience if faced suddenly by a mad dog or a beautiful woman (what we now know to be the effects of adrenaline). Yet these apparently involuntary 'symptoms', he noted, are to a degree under our control, since 'we may either encourage or check' them. For example, controlling an emotion such as fear 'is one of the principal offices of fortitude': instead of fleeing, our mind checks our emotions, and we stand and face our fears. And look at the physical effects a simple letter may have, sometimes causing 'the most extraordinary revolutions in our frame': good news may fill us with happiness and raise our spirits, whilst bad news can plunge us into the depths of misery. Both these mental experiences have physical effects. Indeed, just such a letter 'has been known to occasion death by extreme anguish or extreme joy'.[128] Furthermore, mental effort can be used to control pain, and it could conceivably cure certain diseases. (The eighteenth century had not named the 'placebo effect', of course, but contemporaries were aware of something akin to it – Mesmer, for example, had shown the power of mind over matter.) As Godwin shows, it was widely believed that mental processes played a part in the way illnesses might develop. At the very least, it was clear to Godwin that a man who walks twenty miles with a sense of urgent purpose arrives at his destination as unbowed and elated as when he set out. Yet the man who walks the same distance in sorrow will, by contrast, arrive exhausted. Where lies the difference, Godwin suggested, but in their respective *states of mind*?[129]

Given these known relationships between mind and body, Godwin suggested that the former could control the ageing processes in the latter, because ageing results – he argued – from changing mental states. In order to prolong life and youthful vitality, Godwin recommended cultivating positive mental habits such as cheerfulness, since every time our mind 'becomes morbid, vacant and melancholy, a certain period is cut off from the length of our lives'. Added to this were two further Godwinian principles. The first was to possess a precise 'conception' of what one wished for: 'It is not a knowledge of anatomy, but a quiet and steady attention to my symptoms,' he believed, 'that will best enable me to correct the distemper from which they spring.' The second principle was benevolence: 'The soul that perpetually flows with kindness and sympathy, will always be cheerful. The man who is perpetually busied in contemplations of public good, will always be active.'[130] The final, magic ingredient was progress: since mind 'is infinite', its ultimate powers were inexorable. Improvement would build upon improvement. This would lead us, gradually, towards physical immortality. Bacon, Descartes, Hufeland and Haller had all seen mind as important – but none had ever carried it quite so far on its own.

To potential doubters, Godwin pointed to the great progress that had been made in the sciences. Would the 'savage inhabitants' of Homeric Europe have been able to predict what human ingenuity would be capable of at the end of the eighteenth century? The power to predict eclipses, weigh the air, explain 'the phenomena of nature so that no prodigies should remain', and even to measure 'the distance and the size of the heavenly bodies'. These clear achievements, all driven by the human mind, 'would not have appeared to them less wonderful, than if we had told them of the possible discovery of the means of maintaining the human body in perpetual youth and vigour'.[131] What needed to be done to achieve this vision of a still distant human future was to focus more on medicines for the *mind*, not just those for the *body*.

One place to begin was sleep. Godwin called this daily act 'death's image', and considered it 'one of the most conspicuous infirmities of the human frame'.[132] Overcoming our need to sleep was a step towards overcoming the need to die. He also conjectured that, some day, cultivated humans would be able to control their sexual urges, and would cease to procreate. Having 'perhaps' become happily 'immortal', why would they need to? Then, the whole population of the Earth 'will be a people of men, and not of children'. In this happy, enlightened state, there would be no war, no crime, no need for a legal system or government, no disease, anguish or melancholy.[133] Godwin's optimism for the future of human happiness appeared almost boundless, as it was for Condorcet.

§

Political Justice explored numerous topics besides the possibility of physical immortality. As well as attacks on the institution of marriage, it included a rejection of Christian dogma and a withering critique of government and social greed. This vivid, provocative book made Godwin one of the most controversial philosophers in England – 'ranked at once', a contemporary declared, 'among men of the highest genius and attainments'.[134]

Yet Godwin's suggestion, which so echoed Condorcet's, that humans might one day immeasurably prolong their lives, was not always well received. In 1796 the poet Samuel Taylor Coleridge wrote that he was by no means convinced by Godwin's suggestion 'that man may be immortal in this life, and that Death is an act of the Will!!!'[135] Most readers, another critic observed in 1798, would think Godwin's speculation 'so fantastic', and his 'systems so unfounded', as to be 'fit subjects rather for ridicule than refutation'.[136] The most famous refutation was made by the Revd Thomas Malthus (1766–1834). An accomplished mathematician and a young Fellow of Jesus College, Cambridge, Malthus was well versed in these immortalist arguments. His eccentric father, Daniel (1730–1800), had known Rousseau, and agreed with Condorcet and Godwin's ideas on the perfectability of humans. He considered human progress inevitable; but his son disagreed, and in 1798 published his reasons anonymously, in the now famous *Essay on the Principle of Population, as it Affects the Future Improvement of Society*. The first edition bore the telling but oft overlooked subtitle: 'With remarks on the speculations of Mr. Godwin, M. Condorcet, and other writers'.[137] Thomas Malthus perceived fatal mathematical flaws in both men's vision of an ever-continuing improvement leading to physical and social perfection. These flaws, Malthus suggested, would render impossible the prophecy of a society in which everyone lived 'in ease, happiness and comparative leisure'.

Malthus began by pointing out the fallacy in Condorcet's argument that inferred 'an unlimited progress from a partial improvement'. Yes, it was certainly true that English farmers had enhanced their sheep stock by selective breeding – in particular the famous Leicestershire, with its large body, small head and short legs. But the suggestion that this process could be continued indefinitely until a sheep had a head and legs the size of, say, a rat, was a 'palpable . . . absurdity'.[138] There was a *limit* to everything. This was as true of improving domesticated animals as it was of improving humans. Besides, selectively breeding *people* could not be achieved 'without condemning all the bad specimens to celibacy'. This was an improbable ambition.[139]

Malthus considered the *Sketch* more likely to 'amuse' than 'convince' its readers.[140] Condorcet's trust in the indefinite prolongation of life was 'in the highest degree unphilosophical and totally unwarranted by any appearances in the laws of nature'.[141] With an Englishman's sense of order and decorum, Malthus preferred the 'grand and consistent theory' of Newton, where facts were facts, and the universe could be understood according to a rational, reasonable and verifiable system, to what he called 'the wild and eccentric hypotheses' of that Frenchman, Descartes. In the Cartesian system that Malthus saw as influencing Condorcet, given enough time, almost *anything* might come to pass.[142] Such a philosophy was simply meaningless.

Condorcet's philosophical framework was flawed, and supporting evidence was absent. 'With regard to the duration of human life,' Malthus suggested, 'there does not appear to have existed from the earliest stages of the world to the present moment the smallest permanent symptom or indication of increasing prolongation.' Though an Anglican clergyman, Malthus appears to have set no store by Genesis. He did not mention the long lives of the patriarchs, confining his argument to what he called the 'authentic history of man' and to sound philosophical, empirical principles.[143] Man is mortal, 'because the invariable experience of all ages has proved the mortality of those materials of which his visible body is made'.[144] That was all there was to it.

Having demolished Condorcet, Malthus proceeded to tackle Godwin. Population, Malthus explained in what became the most famous argument in his *Essay*, would – if left unchecked – always grow faster than the human ability to feed itself. The inevitable eventual pressure on scarce resources would periodically lead to misery, famine, war and death, promising the 'utter destruction' of what Malthus called Godwin's 'beautiful system of equality'.[145] As to Godwin's suggestion that humans would one day cease to procreate, this was simply 'an unfounded conjecture, unsupported by any philosophical probabilities'. It received short shrift in one of Malthus's briefest chapters.[146] He dwelt longer, however, on what he called 'Mr. Godwin's conjecture respecting the future approach of man towards immortality on earth'. He pointed out that no one had ever doubted 'the near, though mysterious, connection of mind and body'. But (like Condorcet), Godwin was advancing his argument 'from a small and partial effect, to a great and general effect'. A person might indeed, through force of mind, disregard 'slight disorders' of the body such as toothache. But they could hardly ignore 'a high fever, the smallpox, or the plague'!

Furthermore, Malthus wondered, might it not be possible for the mind actually to *undermine* health, rather than to improve it? Suppose, he hypothesized, that 'a medicine could be found to immortalize the body'. Was it not likely that 'the greatest conceivable energy of mind would probably exhaust and destroy the strength of the body'? Whilst a temperate, cheerful mind is certainly favourable to health, great intellectual exertions can actually be very debilitating. Such endless mental stimulants, 'continually applied, instead of tending to immortalize . . . would tend very rapidly to destroy the human frame'.[147]

As he had those of Condorcet, Malthus then dismissed Godwin's predictions, made without basis in experience. It might be presupposed that in the future humans would have eyes in the back of their heads and an extra set of arms: 'I should admit the usefulness of the addition', Malthus acknowledged, but there were no previous indications 'from which I could infer the smallest probability' of it actually happening. So there was 'no more genuine indication that man will become immortal upon earth than that he will have four eyes and four hands'.[148] For Malthus, such hypothesizing was simply unphilosophical and unscientific. There was no evidence in human history that anyone had ever lived longer by the 'operation of their intellect'.

Having dismissed their arguments on philosophical grounds, Malthus proceeded to condemn Condorcet and Godwin – for both were sceptics – on religious ones. They only placed such hope in immortal flesh, he explained sanctimoniously, because they had rejected 'eternal life in another state' – the posthumous immortal Christian soul.[149] A rationalist in everything except religion, Malthus was no Romantic. One cannot picture the idealistic young poet Percy Bysshe Shelley knocking as an admiring disciple at Thomas Malthus's door and then running off with his young daughter. He knocked at William Godwin's door, and ran off with *his* young daughter.

§

Figure 3.3 Mary Wollstonecraft Shelley: oil painting by Samuel John. Mary Shelley was only twenty-one when her masterpiece, *Frankenstein*, was published in 1818. Her creation has been described by Anne Kostelanetz Mellor as 'an attempt to achieve the final perfecting of Rousseau's natural man, to produce an immortal being of great physical strength and powerful passions who transcends the chains of social oppression and death' (*Mary Shelley: Her Life, Her Fiction, Her Monsters*, London: Routledge, 1988, 81).

Mary Shelley (1797–1851), the only child of William Godwin and his remarkable wife, Mary Wollstonecraft, draws many of this chapter's themes to an appropriate conclusion. Wollstonecraft had been in Paris at the height of the Revolution, and before marrying Godwin had attempted suicide by throwing herself in the Thames, only to be saved by the Royal Humane Society. She died having given birth to her daughter, Mary, who famously went on to write one of the classics of Romantic gothic horror: *Frankenstein, or The Modern Prometheus* (1818), a book she dedicated to her father.

Mary Shelley would later recall how the idea for *Frankenstein* was stimulated by the philosophical conversations between her husband and his friend, the poet Lord Byron. They had discussed the ideas of Dr Erasmus Darwin (1731–1802) on the generation of life, and pondered whether a corpse might be 'reanimated'. Galvanism, they had mused, 'had given token of such things', and 'perhaps the component parts of a creature might be manufactured, brought together, and endued with vital warmth'?[150] *Frankenstein* also reflects some of her father's interests. In 1799, Godwin followed up the immortalist ideas expressed in *Political Justice* in an historical novel, *St Leon* – in which his sixteenth-century hero discovers the philosopher's stone. But immortal life comes at a price, and what St Leon gains in wealth and endless time on earth he pays for in emotional isolation and unhappiness.

The hero of Mary Shelley's novel, Victor Frankenstein, is similarly seeking a solution to the problem of life and death. He has spent his youth imbibing the works of the sixteenth-century alchemists – in particular Cornelius Agrippa and Paracelsus – and searching after the philosopher's stone. But when he arrives at university, his professor tells him he has been wasting his time. These 'ancient teachers' had 'promised impossibilities, and performed nothing'. But, whilst the 'modern masters' had promised 'very little', they 'have indeed performed miracles . . . They have acquired new and almost unlimited powers.' These were the men to be followed.

To understand life, Frankenstein realizes, 'we must first have recourse to death'. He studies anatomy and labours in tombs and charnel houses, examining 'the natural decay and corruption of the human body'. Life and death, for Frankenstein, are 'ideal bounds' to be transcended; his dreamed-for success in discovering the life-giving force would pour 'a torrent of light into our dark world'. Frankenstein frequents dissecting rooms and slaughterhouses; in a solitary chamber at the top of his house he works tirelessly to find a way of reanimating dead body parts. Aided (as Shelley later phrases it) by 'the working of some powerful engine',[151] the doctor finally infuses 'the spark of life' into his inert creation. But the living corpse that looms before him is a horrifying abomination: and what follows is the famous story of an imperfect, unloved creation that turns violently against its appalled and disgusted creator. Hoping to create life, Frankenstein unleashes death and destruction.

Mary Shelley went on to write a number of short stories that explored further the theme of life and death. She had good cause to ponder over these subjects: her mother had died in giving birth to her; her half-sister and her husband's first wife both committed suicide; her handsome young husband drowned; three of her children died in infancy; and her half-brother died before he was thirty. Her father lived on, however, dying shortly after his eightieth birthday, the sun finally setting on what William Hazlitt described as 'the serene twilight' of Godwin's 'doubtful immortality'.[152]

One of Mary's stories reports on the discovery of an Englishman named Roger Dodsworth – who had been caught in an avalanche and frozen, back in 1654. His vital forces, Shelley explains, 'had been suspended by the action of the frost'. Like Bacon's fly trapped in amber, Dodsworth's 'icy shroud' has preserved him untainted by temporal forces, and he is brought back to life.[153]

Another story, 'The mortal immortal', begins on 16 July 1833, with Shelley's character celebrating his 323rd birthday. 'Am I, then, immortal?' he wonders. It is not electricity or ice, however, that has thus prolonged this man's life: it is the 'immortal elixir' prepared by his former master, Cornelius Agrippa. The reluctant immortal has suffered the sadness of watching his wife grow old and die, whilst he remains as a youth of twenty, continuing down the path of a tedious, wearisome and lonely eternity. He resolves upon suicide, and plans to end his life at the frozen pole. No one, it seems, could welcome immortality. The quest was a Romantic error:

> I yield this body, too tenacious a cage for a soul which thirsts for freedom, to the destructive elements of air and water . . . and, by scattering and annihilating the atoms that compose my frame, set at liberty the life imprisoned within, and so cruelly prevented from soaring from this dim earth to a sphere more congenial to its immortal essence.[154]

❂

CHAPTER 4

From Regeneration to Degeneration

These bodies which now we wear belong to the lower animals; our minds have already outgrown them; already we look upon them with contempt. A time will come when Science will transform them by means which we cannot conjecture, and which, even if explained to us, we could not now understand, just as the savage cannot understand electricity, magnetism, steam. Disease will be extirpated; the causes of decay will be removed; immortality will be invented.

Winwood Reade, *The Martyrdom of Man* (1872)[1]

Wiltshire, south–west England, the 1870s A tall, lank young man, a writer, a naturalist, a free spirit, walks high among the prehistoric earthworks. These grassy downs are an ancient landscape, he thinks to himself, 'alive with the dead'. He has abandoned himself to the glory of the day: the golden corn ripening in the sun, the rising song of skylarks. These are the things he loves to observe and reflect on. He feels like the spirit of the man whose bones lie within the neighbouring burial mound: he is at one with nature and the natural world; his soul has sprung beyond the realm of his mortal body. He has an idea of a greater world, a finer, miracle-like existence beyond life. He would like, somehow, to record this profound spiritual emotion in words. But how could it be done?

'Listening to the sighing of the grass', he later recalls, 'I felt immortality as I felt the beauty of the summer morning, and I thought beyond immortality, of other conditions, more beautiful than existence, higher than immortality.'

This, of course, was the sublime immortality of the human soul – but Richard Jefferies sees, *experiences*, something more than this. It troubles his mind, this mysterious thing; he feels impelled to record it. Journalism, novels for adults and for children – he has written all these. But this profound sensation of illimitable existence demands something more. It will be an autobiography of thoughts, springing from deep within what he reluctantly calls his soul. He will name it *The Story of My Heart*. It is one of the more remarkable – if also one of the more overlooked – books published in Victorian England.

A socialist, a pastoralist and an autodidact, Jefferies sees that modern life is rubbish: stinking, smoking, hectic London and the pursuit of money and possessions have nothing of the vast, grassy downland of Wiltshire. Our sick and feeble bodies do not compare with those of the ancient Spartans, famous as 'the finest race of men' and 'the most beautiful women'. Should it not be possible in this modern age, he

Figure 4.1 Richard Jefferies: photograph. Richard Jefferies's idiosyncratic autobiography, *The Story of My Heart*, captured in vivid language a Victorian atheist's yearning for a life beyond the material, monetary world. He envisaged a future in which 'the effort of all the races of the Earth' would be concentrated 'upon man's body, that it might reach an ideal of shape, and health, and happiness'.

wonders, to replicate such Greek achievement? Should it not be possible to redirect our priorities and create a more perfect human form? 'No science of modern times has yet discovered a plan', Jefferies complains in *The Story of My Heart*, 'to meet the requirements of the millions who live now, no plan by which they might attain similar physical proportion. Some increase of longevity, some slight improvement in the general health is promised, and these are great things, but far, far beneath the ideal.'[2]

A whole new way of thinking about human life is required: the traditional ways of doing things need to be stripped away, and 'the effort of all the races of the earth' needs to concentrate instead 'upon man's body, that it might reach an ideal of shape, and health, and happiness'.[3] Jefferies rejects both new theories of evolution and old religious ideas of divine design. There is no God – at least, not in the way Christianity supposes: in Jefferies's opinion, the human frame shows few signs suggestive of a divine creator. Life is random, merciless: death strikes with cruelty; there is no intelligent overseer, 'not the least trace of directing intelligence in human affairs'. The how and the why of existence remain unknown. With God's removal from the picture, the enormity of existence, nature, the mind, is such that whole new ideas are needed to encompass and understand them.[4] If humans want fitter, healthier bodies and a better, longer life, they must search for them themselves. No help will come from 'God'.[5]

For Jefferies, it is 'perfectly certain that all diseases without exception are preventable'. He sees it as 'absolutely incontrovertible that the ideal shape of the human being is attainable to the exclusion of deformities. It is incontrovertible that there is no necessity for any man to die but of old age.' But Jefferies does not believe that anyone actually *dies* of old age. No. They die of disease or weakness, 'either in themselves or in their ancestors':

> No such thing as old age is known to us. We do not even know what old age would be like, because no one ever lives to it.
>
> Our bodies are full of unsuspected flaws, handed down it may be for thousands of years, and it is of these that we die, and not of natural decay. Till these are eliminated, or as nearly eliminated as possible, we shall never even know what true old age is like, nor what the true natural limit of human life is.

Jefferies reckons that the present 'utmost limit' of human existence is about 105 years: but that is in an 'unnatural state'. The *true* old age of man living in a *natural* state, he believes, 'would be free from very many, if not all, of the petty miseries which now render extreme old age a doubtful blessing'. The limbs would not weaken, the eyes would not fail nor teeth decay, the mind would not lose its memory.[6] Indeed, Jefferies wonders if in 'natural' circumstances death might not, in fact, be 'wholly preventable': 'if the entire human race were united in their efforts to eliminate causes of decay', might not death 'be altogether eliminated'?

The blame for our present inadequateness is to be placed on degeneration: 'we are murdered by our ancestors', Jefferies complains. 'Their dead hands stretch forth from the tomb and drag us down to their mouldering bones.' And we 'in our turn are now at this moment preparing death' for our unborn progeny'.

Jefferies believes in progress – but he is surprised by how *little* progress human civilization has made in the last 12,000 years. Too many are still sick and poor. Whilst the human race is easily capable of creating immense armies for its own destruction, others live hand to mouth, in poverty, in a superabundant world. The reasons, he declares, are selfishness, lack of organization and the effects 'of an age infatuated with money'. All the efforts of past ages have been wasted; the millions of people rushing to and fro in the banks and the stock exchange and the shops and factories of London have accumulated nothing useful: 'The only things that have been stored up have been for our evil and destruction, diseases and weaknesses crossed and cultivated and rendered almost part and parcel of our very bones.'

We work like slaves, yet there is such plenty on earth that one day, with better organization, and by dividing and sharing, it should be possible for people to spend nine tenths of their time in idleness, enjoying the beauties of nature. We possess 'infinite' minds with which to think, and hands with which to create; we possess a soul that, though currently in abeyance, 'may yet discover things now deemed supernatural'. The human being, Jefferies asserts, 'is shaped for a species of physical immortality'. With 'united effort' continued over the course of long ages, the causes of decay may be eliminated, and the improvement of the 'natural state', advanced without end. It is time for a change, 'to roll back the tide of death, and to set our faces steadily to a future of life'.[7]

This is not a scientific scheme: it is Jefferies's personal realization of something that exists beyond body and soul, of 'the immensity of thought which lies outside the knowledge of the senses'.[8] The knowledge of facts is limitless; they lie at our feet like innumerable, countless pebbles. But the mind is currently circumscribed, the same predictable thoughts have carried on down the centuries. It is time, Jefferies suggests, to break through those boundaries and think *differently*, to explore *new* ideas.

He is no scientist; nor is he, in truth, a philosopher. From a humble background of Wiltshire farmers and London printers, Jefferies had left school at fifteen. Through force of will he carved a career as a journalist, a novelist, a mystic, a visionary – the 'prophet of an age not yet come into being', as one biographer puts it.[9] In one of his most famous novels, *After London*, the capital of the British Empire is submerged beneath a vast swamp, people return to barbarism, and the countryside reverts to what he calls 'wild England'. *The Story of my Heart*, published when he is thirty-four, does not sell well: a fellow writer teases him on the 'uselessness' of his pursuing such philosophical musings. But Jefferies stands by all he has written there; he claims his book has found adherents in 'far-flung places'.[10]

Yet all is not well with Jefferies's health as he writes his autobiography of thoughts: though still a young man, he is ailing. It's a slow, wasting disorder that baffles his doctors. In 1885 his youngest son, a child of less than two years old, dies. Jefferies is too distraught to attend the boy's funeral. By June that year, he is fainting day and night. He craves food yet is almost unable to eat. He is, he writes, 'half delirious, and in the most dreadful state', his nerves 'gone to pieces'.

In September his spine seems suddenly to snap: 'It felt as if one of the vertebrae had been taken away', he tells a friend. He is left scarcely able to walk. Unsteady, he becomes weaker, emaciated, immobile. 'Often I am compelled to sit or lie for days and think, think,' he writes, 'till I feel as if I should become insane, for my mind seems as clear as ever . . . There is an ancient story of a living man tied to a dead one, and that is like me; mind alive and body dead.' One physician suggests it is tuberculosis of the lungs – or an unusual case of male hysteria. Others, similarly, suggest there is – physically speaking – *nothing* wrong with him: that he is suffering from nervous exhaustion rather than disease, the effects of his ceaseless work, the ominous London air, the sudden death of his son. The likely diagnosis is chronic fibroid phthisis – a form of tuberculosis, and one of the period's biggest killers, accounting for around one in four deaths in nineteenth-century England.

After vomiting blood, Jefferies moves with his family to Goring-on-Sea, in the hope that the bracing winter air of the Sussex coast will restore his collapsing health. In January 1887 he writes in his notebook that he is 'half-paralysed in the chilly frost & damp – like a crawling caterpillar'. No one, he reflects, looking now at his 'decrepit form', slobbering his food 'like an imbecile old man', unable to walk even a hundred yards in the world he loves, would ever imagine he had once been plotting physical human perfection. '*Science*', he records, cynically, ambiguously, 'cannot cure a disease more now than then.'[11]

'I burn life like a torch', he had written in *The Story of My Heart*.[12] At half past two on the morning of 14/15 August, his body 'wasted to a skeleton by six long weary years of illness', Richard Jefferies's brilliant light gutters out. He is only thirty-eight years old.[13]

§

Jefferies's reflections on the perfectibility of the human species and the prolongation of life illuminate two key ideas with a long, slow historical gestation which finally bore fruit in the nineteenth century: evolution and the death of God. The latter contains within its bounds that subject I have largely and deliberately avoided: the immortality of the soul. It cannot be ignored any longer, and I will explore it briefly, together with the diminishing authority of the Bible over this period. Then I can return at length to the question of evolution, and the nineteenth century's dream of 'perfecting' the human species.

Philosophical and theological arguments in favour of the soul's immortality had come under increasing pressure since the late seventeenth century – particularly in England. The pressure gathered force in the eighteenth. John Locke had mounted one of the earliest and strongest critiques, whilst in 1777 the Scottish Enlightenment philosopher David Hume asked what arguments or analogies could prove the existence of something which nobody had ever observed, and which resembled nothing that had ever been seen. 'Who will repose such trust in any pretended philosophy', he questioned in his essay

'The immortality of the soul', 'as to admit upon its testimony the reality of so marvellous a scene?' Only some 'new species of logic' or 'some new faculties of the mind' could prove this great unknown.[14]

Yet, if the soul was *not* immortal, if there was no afterlife, what exactly was the point of our mundane existence? Man's very being, if it was to have a purpose, seemed to stand on the promise and possibilities, not of this life, but of an eternal afterlife beyond the grave. William Wollaston speculated in 1722 whether it was really conceivable that the human being could 'be designed for *nothing further*, than just to eat, drink, sleep, walk about, and act upon this Earth', with no greater purpose than the sheep and cows in the fields. 'Can I be made capable of such *great expectations*', he pondered, 'only to be *disappointed at last?*'[15] Henry Grove, writing in the *Spectator* in 1714, also saw the seeming pointlessness of everyday life, contrasted with our constant striving for new experiences, as clear *proof* of the soul's immortality. The very vanity and vacuousness of life seemed to indicate to Grove that (spiritually at least) the human being '*is designed for Immortality*'.[16]

Wollaston and Grove were hardly alone in their thoughts, but some took them to atheistical conclusions. Who can be sure that the reason for man's existence, pondered the French physician Julien de la Mettrie in his 1748 essay *Man a Machine*, 'was not simply the fact that he existed?' 'Perhaps he was thrown by chance on some spot of the surface of the earth', he mused, 'without a possibility of discovering why or whence he came; and with this knowledge only, that he must live and die, like the mushrooms which appear from day to day, or like those flowers which border the ditches and cover the walls.'[17]

What if there was no soul, no afterlife? What if life really *was* purposeless? To some – many, probably – the possibility was horrifying. 'If there is no immortality,' declared the Victorian poet Alfred, Lord Tennyson, 'I shall hurl myself into the sea.'[18] By Dover beach, another poet, Matthew Arnold, heard the tide of Christian faith retreating from the wide world's shore with a 'melancholy, long, withdrawing roar'. He was filled with dread at the thought of life in an uncertain, loveless, painful world, 'Swept with confused alarms of struggle and flight, Where ignorant armies clash by night'.[19]

Yet atheism continued its advance. The German philosopher Friedrich Nietzsche did not fear staring into the abyss: God is Dead, he declared in the 1880s; all that remained for Nietzsche in this theological vacuum was man and superman. As Jefferies had counselled, we had to look within ourselves if we were to depend upon ourselves.

Deepening criticisms of the Bible paralleled the attacks on the doctrine of the immortal soul. In the seventeenth century Thomas Hobbes had questioned the authenticity of various parts of the Pentateuch – the first five books of the Bible, including Genesis, which were traditionally attributed to the pen of Moses. Whilst the French Protestant scholar Isaac de La Peyrère speculated whether there had been races of men created *before* Adam, the Catholic Church tightened its rein on such speculations: in 1600 the philosopher Giordano Bruno was burnt in Rome for various heresies, including the suggestion that humans were not the descendants of two Edenic parents. A few decades later, the Italian astronomer Galileo Galilei was placed under house arrest for his defence of Copernicus's suggestion that the earth circled the sun. The Inquisition ruled Catholic Europe with a rod of red-hot iron.

But, as closer scholarly exegesis of the Bible advanced in Protestant nations, criticism of its authority and authenticity became ever stronger. Certainly, numerous scholars argued, biblical claims for the longevity of the patriarchs were not to be taken literally: the Fall was simply an origin myth, like those found in the histories of other civilizations. Samuel Johnson's friend Hester Thrale was shocked when in 1790 her local vicar remarked 'that Adam & Eve and the Apple was an old Woman's Story'.[20]

By the nineteenth century, God – and the Bible – were no longer sufficient to provide answers for understanding what was increasingly recognized as a long, complex human history. The divine architect was taking a hammering, the edifice of his Church falling away in tiny, continuous pieces. Scientists active in the new discipline of geology could no longer accept biblical arguments to the effect that the world was only a few thousand years old. The evidence from fossils, and the gradual changes seemingly wrought by the elements on the face and foundations of the earth, suggested that the latter was in fact many thousands – if not hundreds of thousands – of years old. Theorists such as Hooke, Buffon and Lamark threw Archbishop Ussher's meticulous, seventeenth-century dating of creation to 4004 BC out of the window. Among what Hester Thrale considered the 'many other equally *impious* Positions' expressed by her freethinking vicar in 1790 was his belief that 'the World was five Hundred thousand years old at least'. It was all too much for Mrs Thrale, who thought the end of the world must come soon.[21]

§

The Enlightenment philosophy of Buffon, Hufeland and Haller had heralded an increasingly scientific, less prosaic approach to understanding human ageing. However, compared with the previous two centuries and with the highpoint of speculation that occurred in the 1790s, the nineteenth century generally showed a marked uncertainty at the possibility of vastly prolonging human life.

Early in the century, the wealthy Scottish politician Sir John Sinclair (1754–1835) published numerous editions of his multi-volume *Code of Health and Longevity: Or, A General View of the Rules and Principles Calculated for the Preservation of Health, and the Attainment of Long Life*. Sinclair – who, although well educated and highly inquisitive, had no medical experience – saw no evidence that the human frame had been built to last for a long time. Rather, he observed that the human body clearly 'cannot possibly retain life for ever'. This was because of the gradual, inevitable hardening and rigidity of the parts, such as the blood vessels, which eventually become 'incapable of executing their offices'. He cited Sir Francis Bacon's *History of Life and Death* in support of this argument, and Bacon's suggestion that almost all the original parts of the body, including the sinews, arteries, bones and bowels, 'are hardly reparable'. Ignoring Bacon's suggestion that physical immortality *might* actually be feasible, Sinclair concluded that 'nothing' was 'more absurd than the idea that the human frame can be preserved for ever, or even for a much longer period than at present'. Sinclair felt it was necessary to impress this fact on his readers, since, he noted, 'not only quacks, like Paracelsus, but even distinguished philosophers, like Descartes, have imagined it possible to prolong life very considerably beyond the common period'.[22]

Yet even Sinclair believed that contemporary lifespans were not what they had once been. He noted Hufeland's observation that in patriarchal times a year was measured as three months, not twelve: this would make Methusaleh a more believable 240 years old.[23] Inevitably, Sinclair's main piece of evidence was the one we have found to be repeated over and over again: what he called the first known 'anatomical account drawn up of the dissection of any old person' – that is, William Harvey's seventeenth-century dissection of Thomas Parr. Sinclair blamed the shorter lives of modern times on the rapidity with which people were expected to grow up: the poor, he explained, were forced to go out and work too young, whilst the children of the rich had their education hastened and were rushed too soon into the adult world. He cited Hufeland's example of Louis II, King of Hungary, who, 'it is said, was born so long before the natural time, that he had no skin; in his second year he was crowned; in his tenth year he succeeded; in his fourteenth year he had a complete beard; in his fifteenth he married; in his eighteenth he had grey hairs; and in his twentieth he died'.[24] Such a life was lived too fast to be lived long.

In his 1815 book on the effects of regimen in treating cancer, William Lambe (1765–1847), a Fellow of the Royal College of Physicians, observed that examples of 'extraordinary longevity' showed 'how much we are in the dark on these subjects. Men have arrived at double, and more than double, of what is the greatest common extent of human life.' For Lambe, the 'real wonder' was 'that such multitudes perish prematurely'.[25] This appears to have been a common trope: in 1837, *Chamber's Edinburgh Journal* affirmed that the fact 'that some men have attained an age beyond 150 years in length' was 'sufficient to prove that the human frame is not formed for only a short term of existence, and that much may reasonably be expected from attention to the laws which regulate the health'.[26] The anomaly, it appeared, was not in those few who lived long, but rather in those many who did not.

As in earlier centuries, therefore, Lambe blamed 'excess' and 'luxury' for making most people weak and comparatively short-lived. Over time, the accumulated effects of eating meat 'and other noxious matter' had induced and accelerated 'fatal disease'. This had engendered weakened constitutions, had rendered many people's lives 'a long continued sickness', and had made 'the great mass of society morbidly susceptible of many passing impressions, which would have no injurious influence upon healthy systems'.[27] Lambe was a strict vegetarian from his early forties, and his grandson later recalled how the old man's unexpected appearance at breakfast would produce 'great commotion' as the table was swiftly cleared of meat, with any stray pieces 'heaped on the plate of a carnivorous governess, who, being ignorant of the reason, looked on in bewildered amazement'.[28] Lambe, who lived on into his early eighties, interpreted disease as the product of 'the wastefulness and prodigality of man'.[29]

Excess was one effect of capitalism. It had other evils. In the early 1830s Asiatic cholera arrived in Europe, and the stinking, overcrowded cities, with their filthy water supplies, were partly to blame for this deadly epidemic. They had, it was claimed, bred a degenerated, impoverished, malnourished and overworked underclass. 'The growth of civilization', *The Times* observed in 1868, 'means the growth of towns and the growth of towns means, at present, a terrible sacrifice of human life . . . The fact is that in creating towns, men create the materials for an immense hotbed of disease, and this effect can only be neutralized by extraordinary artificial precautions.'[30]

Guides for health therefore remained popular, catering for the emergent, health-conscious middle class. But they did not bring much that was new to the debate. Some were positively old-fashioned. Dr James Johnson (1777–1845), a former naval surgeon, had firm ideas on the nature of life and death. He appended an essay on the prolongation of life to his 1818 treatise on disorders of the digestive and nervous systems, in which he explored the six Hippocratic 'non-naturals' and their effects on health.[31] In 1836 he followed this work with his popular study, *The Economy of Health, Or the Stream of Human Life from the Cradle to the Grave*. There he observed that the study of the human frame led 'not unnaturally . . . to the conjecture, that MAN was *designed* for immortality, when first turned out of his Creator's hands'. It was only 'melancholy experience', he explained, that shows us that either this 'design of immortality was abandoned by the divine Architect', or that some 'mysterious and fatal revolution' had occurred 'in the constitution of mankind'.[32]

A second edition of Johnson's book appeared the following year, and it had reached a fourth edition by 1843. Johnson commented in a footnote that, though this passage on human immortality had been 'condemned' by one of his critics, he was 'still of opinion that the conjecture is not unnatural'.[33] Notably, however, Johnson believed that physical immortality, 'or even a considerable prolongation of man's existence in this world', would be 'the greatest curse' God could inflict on humans. Life was already long enough. Any more of it, and boredom would prevail: even before this brief span on earth was completed, Johnson suggested, 'the appetite for pleasure begins to be sated . . . and were his years tripled or quadrupled, this earth would fail to afford novelty, and sameness of scene would sicken every sense! If a MILLENNIUM should ever obtain in this world, there must first be a new creation of beings, and that of a nature by us totally inconceivable.'[34]

Fundamentally, ideas on what ageing was and on how it happened changed little over the century. In 1876, *Chamber's Journal of Popular Literature, Science and Arts* set it out in terms that would not have been too unfamiliar to Descartes or Boyle:

> Continually, and as long as life lasts, the particles of our bodies are being disintegrated and cast off, as the result of the wear and tear of vital machinery. The active stream of the blood circulation is ever carrying away to the lungs and other excretory organs, the useless particles it has received from the tissues; whilst this same vital stream is as continually bringing new particles to replace the old. Thus, the body resembles a house which is being gradually taken to pieces, whilst at the same time it is being incessantly repaired.

In our period of growth, repair exceeds wear; then for a time they are equally balanced. But in old age we wear out faster than we repair, 'and the edifice of poor humanity at last succumbs, for want of material and energy to maintain it'.[35]

Some believed that the future would hold the answer to man's mortality. In 1873 George Harris, Vice-President of the Anthropological Institute of Great Britain and Ireland, was speculating on the possibility that new medicines would one day infinitely prolong human life. 'Is it altogether irrational,' he asked in the Institute's journal, 'to suppose that some principle analogous to that of vaccination, or to that supposed to be contained in the very tree of life itself, may at some distant period in the progress of science be brought to light by which the animal frame may be reinvigorated and rescued from decay, and so fitted to endure, I will not presume to say for ever, but to an age corresponding with that to which we are told that both the patriarchs and many animals have attained?'[36]

In *Longevity: The Means of Prolonging Life after Middle Age* (1874), Dr John Gardner (1804–80) pondered the possibility of some day 'preserving the integrity of the body and all its parts with the vigour of youth and middle-age up to the extreme limit of life'. This was not an idle question, he asserted: 'Scientific ideas run in long grooves', and discoveries such as anaesthetics, inoculation and vaccination, as well as more recent ones in organic chemistry, 'greatly encouraged' him 'to hope for the discovery, at no distant period, of an agent which shall effectually stay the changes in the system incident to age – in short, arrest the progress of what we call decay'. Gardner, who had been instrumental in the foundation of the Royal College of Chemistry, felt that, if only the general public 'could be brought to apprehend the prospect of success' of finding scientific means to prolong life, the 'pecuniary aid' necessary to fund the research would certainly be forthcoming.

A devout Christian, Gardner asserted that there was 'no fact reached by science' that contradicted the 'extraordinary longevity' of the patriarchs. 'It is more difficult, on scientific grounds,' he wrote, 'to explain why men die at all, than to believe in the duration of life for one thousand years.' This faith-based belief appears to have played a part in Gardner's conviction that human life could be substantially prolonged. 'At some future time,' he concluded, somewhat disappointedly, 'the indifference and neglect of the present will excite remark and wonder.'[37]

§

Even if a new medicine to prolong life was not on the immediate horizon, statistics seemed to support optimistic interpretations of human progress. By 1826 life expectancy at birth in England has risen to 41.3 years, a vast improvement upon the 25.3 years that could be expected in 1726. Though by 1850 it would have dropped again, to 39.5 years, the general tendency was upwards.[38] A review of John Harrison Curtis's *Observations on the Preservation of Health, in Infancy, Youth, Manhood, and Age*, published in *Chamber's Edinburgh Journal* in 1837, reported that it was 'a vulgar error to suppose that men are less strong and vigorous, and therefore shorter lived, now than in former times'. The reviewer was particularly impressed to note how recent studies had shown that 'the average term of existence' had actually increased.[39] As Charles Dickens observed in his 1854 novel, *Hard Times*, 'the average duration of human life is proved to have increased of late years. The calculations of various life assurance and annuity offices, among other figures which cannot go wrong, have established the fact.'[40]

The science of studying human mortality was undoubtedly improving. John Graunt and Edmond Halley had used Bills of Mortality to reach their conclusions on human life expectancy in the seventeenth century. Of increasing interest from the early eighteenth century on were life annuities. These are known to have existed since the thirteenth century, and in the sixteenth century Amsterdam and Venice used them as a way of raising funds: essentially, a life annuity works in the form of an individual (or, in the case of tontines, of a group of subscribers) paying an upfront lump sum (effectively a loan). In return, they receive a set annual fee from the recipient of the loan for the remainder of their life: these annuities can be seen as an early form of pension planning. It is, in effect, a gamble: the recipient of the lump sum hopes the donor will die before the full sum of the loan has been paid back, whilst the recipient of the annuity hopes that they will outlive the value of the loan.

Life annuities were first used by central governments in the late seventeenth century as a way of raising money, first in France, and then in England. But it was soon realized that, given accurate life tables of the sort Graunt and Halley had studied, life annuities could be mathematically modelled, and the statistical probability of an individual living a certain length of time could be calculated.

The mathematician Abraham de Moivre (1667–1754) made an important advance in this new science. A French Huguenot who fled to England in the 1680s and was influenced by Newton's discoveries in calculus, de Moivre published in 1718 *Doctrine of Chances*. His *Annuities on Lives* followed in 1725 – and both books went through a number of important revised editions. Notably, de Moivre reckoned that nobody lived beyond eighty-six years. This is what the most reliable evidence told him, and in the preface to *Annuities on Lives* he explained that he was 'no more moved' to change his mind by some recent observations 'that Life is carried to 90, 95, and even to 100 years', than he was 'by the Examples of *Parr*, or *Jenkins*'.[41]

Statistics would lead towards a proper appreciation of the length of human life. This happened partly because longevity became crucial from a financial point of view – get it *right* and price an annuity correctly, it was a way of making money; get it *wrong*, and it was an easy way of losing it. Statistical studies such as those contained in the French mathematician Antoine Deparcieux's *Essay on the Probabilities of the Duration of Human Life* (1746) were thus important steps in understanding how individual lives should be correctly valued. Statistical studies would also confirm commonly held assumptions such as that people lived longer in the countryside than in cities. It was true: they did.

In turn, as the historical records of subscribers to life annuities grew, it became increasingly possible to calculate mortality statistics at a given age, and for a given gender.[42] It is interesting to note that, at first, it was governments that tended to get it wrong, and investors who got it right: this was partly because governments offered a flat rate on annuities to all comers (and to those who wanted to invest in the lives of others), regardless of whether they were eighteen or eighty, man or woman, healthy or unhealthy; and partly because the mortality tables they worked from were inaccurate. Whilst the British government's actuary reckoned that only one male in 138 reached the age of ninety nationally in 1831, in Northumberland it appeared that one in 44 reached that grand age. The poet William Wordsworth was among those who made money by investing in badly undervalued government annuities on the lives of healthy rural octogenarians.[43]

As well as life annuities, life assurance was also becoming increasingly a big business, and generated the same interest in compiling tables of mortality. In 1765, the philosopher Richard Price was asked to assist the recently founded Society for Equitable Assurances to compute rates for annuities. Price made detailed calculations based on the inhabitants of Northampton, a small market town in the English Midlands, which would be used for over a century. The tables constructed by the Sun Life Assurance Society's actuary Joshua Milne, using the Bills of Mortality for Carlisle, improved on these. They were themselves bettered in 1823 by John Finlaison, who later became actuary to the British government.

Such researches and calculations on ageing and death seemed to make it clear that it should be possible to establish a *law* of human mortality. The English mathematician and actuary Benjamin Gompertz (1779–1865) achieved this in the early 1820s.[44] Gompertz realized that what he called a 'law of geometrical progression' pervaded 'large portions of different tables of mortality', and that this could actually be expressed in a simple equation.[45] He published his work, based on the Carlisle and

Northampton tables and on Deparcieux's observations, in an important essay that appeared in the Royal Society's *Philosophical Transactions* in 1825. It established what, with further revisions, would become known as the Gompertz law: in essence, Gompertz's equation showed that, from sexual maturity to old age, human death rates rise exponentially. Put in its modern form, for every eight years that we live, our statistical chance of dying doubles.[46]

But why is this so? Gompertz suggested that death was 'the consequence of two generally co-existing causes'. Plain chance was one. The other was what he called 'deterioration, or an increased inability to withstand destruction'. The root of the latter, the 'increased liability to death' as we get older, is due to the fact that mankind is 'continually gaining seeds of indisposition' – something we will return to later in this chapter.

Significantly, Gompertz suggested that the law of mortality he had identified 'would indeed make it appear that there was *no positive limit to a person's age*' (my italics). But Gompertz added that 'it would be easy . . . to show that a very limited age might be assumed to which it would be extremely improbable that any one should have been known to attain'. From Milne's Carlisle tables, he saw that, in the age-range of ninety-two to ninety-nine, one in four alive at the commencement of each year was dead by the end of it. Thus the odds would be above a million to one 'that out of three million persons, whom history might name to have reached the age of 92, not one would have attained to the age of 192'. Nevertheless, though the projection of his equation makes it *highly unlikely* that anyone would live to attain nearly two centuries, it did not make it *impossible*: whilst a quarter of the number may die each year, although this number would eventually fall below one, as a fraction of a fraction it would *never* actually reach zero.

The 'limit to the possible duration of life', Gompertz thus pointed out, 'is a subject not likely ever to be determined, even should it exist'. Indeed, he observed in his *Philosophical Transactions* article that 'the non-appearance on the page of history of a single circumstance of a person having arrived at a certain limited age, would not be the least proof of a limit of the age of man'. Furthermore, 'neither profane history nor modern experience could contradict the possibility of the great age of the patriarchs of the scripture'. Indeed, 'if any argument can be adduced to prove the necessary termination of life, it does not appear likely that the materials for such can in strict logic be gathered from the relation of history'.[47] So, although mortality statistics showed the increasing likelihood of death the older we become, they could never conclusively prove that an absolute age-limit existed and had to be reached.

For Sir Henry Holland, physician-in-ordinary to Queen Victoria and author of an influential essay on longevity published in the *Edinburgh Review* in January 1857, the information gathered by life assurance companies, together with that from the national decennial census that had started in 1801, were 'our best guides' to exactly how long humans might live.[48] Holland's essay was prompted by a book written by Pierre Marie Flourens, perpetual secretary to the Academy of Science and Professor of Comparative Physiology at the museum of natural history in Paris. In *On Human Longevity and the Amount of Life upon the Globe* (1854; English translation, 1855), Flourens had reasserted Buffon's claim that one hundred years was our maximum natural lifespan. But he had added that it was a 'fact' or 'law' that in 'extraordinary' circumstances that span could be prolonged to at least 150 years, if not even to two centuries. An 'extreme' example was Thomas Parr's 152 years. This, Flourens noted, was significant 'because it cannot be disputed [as] it has the testimony of Harvey'. An 'extraordinary life' of a century and a half, he asserted, was 'the prospect science holds out to man'.[49]

Holland, however, found Flourens's conclusion 'to be unfounded; and his arguments on behalf of it vague and unsatisfactory'. His instances of long life were 'often of exaggerated or doubtful kind'. Modern statistical records of mortality, Holland pointed out, 'utterly' refuted Flourens's argument. 'A hundred years is not,' Holland asserted, 'and has never been, the natural or normal age of man.'[50] Yet this did not mean that Holland dismissed the idea that some humans could and did live as long as a century. Whilst he felt that the examples of Henry Jenkins and the Countess Desmond were doubtful, he could not reject 'the evidence as to the 152 years of Thomas Parr's life, accredited as it is by the testimony of Harvey'. In fact, Holland observed that there were other instances 'of this extraordinary kind . . . fully admitted by some of the most eminent physiologists'. For Holland, there was 'sufficient proof of the occasional prolongation of life to periods of from 110, to 130, or 140 years'. Such anomalies occurred in human height and weight, he observed, so why not also in age? Indeed, he noted that there were anomalies 'either of excess or deficiency' occurring 'in every part of the physical structure of man', as in the rest of nature.[51] Why should lifespan be any different? What mattered, he explained, was the law of averages:

> The law of averages, indeed, has acquired of late a wonderful extension and generality of use; attaining results, from the progressive multiplication of facts, which are ever more nearly approaching to the fixedness and certainty of mathematical formulæ. Every single observation, and every new fact added, comes into contributing to these resulting truths. Phenomena, seemingly the most insulated, and anomalies the most inexplicable, are thus submitted to laws which control and govern the whole.

Given the law of averages, Holland held that the idea of a 'natural' lifespan was 'an abstract conception incapable of being expressed, otherwise than approximately, by any simple number'. If pushed, Holland would go along with the psalm: three score year and ten was probably about right – though even this was probably optimistic.[52] Life beyond this mark was possible – but it was anomalous.

Here at last we see some progress. But, though it was statisticians who showed that average lifespans were definitely on the increase in Britain, it was not a scientist or an actuary who finally put a maximum known length on human lifespans. As we shall now see, it was an historian.

§

The authority of the long lives of the patriarchs diminished in the nineteenth century. Instead, secular examples such as those of Parr, Jenkins, Czartan and other local figures recurred in the popular press, and even in medical works. Though Abraham de Moivre dismissed Parr's great age out of hand, it convinced Flourens and Holland, and in 1853 Dr Barnard van Oven cited Parr's great age in the preface to *On the Decline of Life in Health and Disease*.[53] Indeed, in extensive appendices, van Oven gave tables with the names, occupation (where known), country of residence and year of death of 1,519 men and women who had reputedly lived between 100 and 110 years; 478 who had lived from 110 to 150 years; and a remarkable 17 who had lived over 150 years. In addition, he listed 54 people over 100 who were *still alive* at the time he was writing his report. These included Margaret Forster, an Englishwoman born in 1771, who was 136.[54] A review of van Oven's book in *Bentley's Miscellany* praised its author, who 'speaks in the spirit of a gentleman and a man of science, and with a mild wisdom'. In a world in which most lives were cut brutally short, the reviewer felt that van Oven's book might 'breathe hope, solace and encouragement to ears that had forgotten their very sound . . . We heartily recommend this extraordinary little work.'[55] A century and a half to two centuries appeared to be the 'natural' lifespan.

By the 1860s, however, this was all getting too much for William Thoms (1803–85), Fellow of the Society of Antiquaries of London, Deputy Librarian at the House of Lords and founder (in 1849) of the esteemed literary journal *Notes and Queries*. Thoms complained of 'the unhesitating confidence and the frequency with which the public is told of instances of persons living to be a hundred years of age and upwards'. This, he felt, 'so familiarises the mind to the belief that Centenarianism is a matter of everyday occurrence, that the idea of questioning the truth of any such statements never appears to have suggested itself'.[56]

Thoms was particularly struck by the 'child-like faith' with which even men of 'the highest eminence in medical science', such as Sir Henry Holland, accepted 'without doubt or hesitation statements of the abnormal prolongation of human life'.[57] It was an article Thoms published in *Notes and Queries* in 1862 that first kindled his interest. It contained details of what appeared to be a perfectly authentic account of Mrs Esther Strike, born at Wingfield, Berkshire, on 3 January 1659, and who had died at Cranbourne St Peter's, Berkshire, on 22 February 1762. The author of the article, Sir George C. Lewis, stated that he had found no other 'well-authenticated case of a life exceeding 100 years', either in the records of modern peerage and baronetage, or in the experience of life insurance companies. These facts suggested to Lewis that human life 'under its ordinary conditions' was never prolonged much beyond a century.[58]

But what of all those commonplace reports of centenarians in the press? It was just such a report that stirred Thoms into action. In 1863,

The Times announced that a Miss Mary Billinge had died in Liverpool on 20 December 1863, aged 112 years. Instead of taking this at face value, Thoms decided to investigate. Whilst what he called 'persons of intelligence and position' accepted that Miss Billinge had indeed died in her 113th year, he took the hitherto almost unprecedented step of actually checking the records for himself.

Though Henry Holland had expressed doubts about many claims of long lives, the only persons who seem to have attempted actually to check the records earlier than Thoms were Sir George Lewis and the Danish civil servant Bolle Luxdorph. Yet baptisms, burials and marriages had been registered in England and Wales since 1537. And, as Thoms's investigation of the appropriate parish records revealed, the Mary Billinge whose death in 1863 had been reported in *The Times* was *not* the same Mary Billinge, daughter of William Billinge, who had been born on the 24 May 1751. She was, in fact, that woman's *niece* – the spinster daughter of the other Mary's brother. This Mary had been born on 6 November 1772. So she was 91 years old, not 113 at all. Old, yes; but still some way off even a century.

Having demolished Mary Billinge's claim, Thoms proceeded to use the same method to try and prove that Mrs Williams of Bridehead, 'believed by her family to have been 102 at the time of her death, on the 8th October 1841', was not in fact so old as claimed. Thoms traced the original record of the old lady's baptism to the parish registers of St Martin-in-the-Fields, London, on 13 November 1739. To his great surprise, he discovered that the old lady really *was* 102! Whilst admitting that he came off 'second best' in this 'friendly controversy', by establishing Mrs Williams's 'exceptional age' he only became even more interested in such enquiries. Thoms went on to accept numerous challenges to investigate similar cases. Eventually they grew so numerous and time-consuming that he published a circular explaining that he was not going to undertake any more.

Thoms published the results of these labours in 1873 as *Human Longevity: Its Facts and Fictions.* He eventually wished he had never bothered. As he observed in the second edition of this book, he had become noted in the press 'as the ardent apostle of the strange scepticism that nobody exists to over 100 years'. He advised anyone who had 'the slightest desire to live in peace and quietness' not 'under any circumstances, to enter upon the chivalrous task of trying to correct a popular error'. The public response to his book had been such, he wrote, that he considered himself 'one of the best abused men in England'.[59]

Thoms's experience shows how deeply ingrained the belief in long life had become. It was almost as if he was personally removing the chance of living to 150. Yet, given 'the number of works which have been written on Old Age, the means of attaining it, and other matters connected with the duration of human life', Thoms was surprised that his book was 'the first in which the important question, What is the average extension of human life? has ever been tried by the logic of facts'.[60] Amongst the many claims he investigated were what he called the three 'stock cases' of Thomas Parr, Henry Jenkins and the Countess of Desmond. For so long, these three had been the cornerstones in the medical debate for the prolongation of human life. But, as Thoms demonstrated, there was not one shred of historical evidence in support of *any* of them.

In fact, the longest proven human life he found, based on marriage and baptismal records, was that of a Martha Lawrence. Born on 9 August 1758, she had died on 17 February 1862, aged 103 years and six months.[61] Although *Chamber's Journal* explained that Thoms's was 'a work which everybody who has not read it should read',[62] many clearly did not. Whilst the medical investigation of old age was improving throughout the later part of the century – particularly with the work of Alfred Loomis in New York and J. M. Charcot in Paris from the 1860s to the 1880s – the press continued to indulge in stories of remarkable human longevity. In 1874, the *Birmingham Morning News* reported the death of a Muslim man in Madras aged 143; in 1879, *The Telegraph* carried the story of a 'venerable gentleman' in Mexico aged 180; and, according to the Registrar General in 1885, at least one person in a million born in England could expect to live to 108.[63] In 1888 John Burn Bailey even published another catalogue of longevity, with real nonagenarians such as Caroline Herschel (1750–1848) mixed in with the spurious. In the preface to his *Modern Methuselahs*, Bailey declared that human life was capable 'of an extension to three times its present duration'.[64]

Yet even those with aims and modest expectations similar to Thoms's could make mistakes. Using well-kept parish records in Canada in the 1870s, the doctor and government statistician Joseph-

Charles Taché seemed to prove that Pierre Joubert, who had died in 1814, was 113 years old. The Pierre Joubert in question, born in 1701, actually died in 1766, and it was his homonymous son who died in 1814. This was a common error in trying to authenticate age at death, but this 'record' was particularly durable: it seems to have appeared in the *Guinness Book of Records* as late as 1997.[65] But this was nothing close to the 150 to 200 years approved of, only a few decades earlier, by Sir Henry Holland, Haller and Hufeland. Thoms's work marked a watershed in the history of longevity, removing an error that had long fooled those interested in the subject.

§

If diet and fresh air could no longer be expected to lead one to a life of 150 years, was there another way? In the 1880s Richard Jefferies wrote of his hope that modern society might some day spawn a race of men and women to rival the apotheosis of Greek strength and beauty – the Spartans. But how did he think this was to be achieved? He did not say – other than by some sort of collective action. But Jefferies grew up in farming, and he knew at least something about selective breeding and the improvement of cattle. It was a subject he wrote about occasionally, for periodicals such as *Live Stock Journal and Fancier's Gazette*. This, perhaps, is where he saw gradual improvement coming from. It is certainly where many nineteenth-century thinkers looked when they sought the improvement of the human race.

As the focus in the later eighteenth century had been on diet and the power of the mind, so in the nineteenth the role of sex, breeding and heredity became paramount. Sir William Lawrence, in his *Lectures on Physiology, Zoology and the Natural History of Man* delivered at the Royal College of Surgeons in London and published in 1822, asserted that a 'superior breed of human beings' could only be produced 'by selections and exclusions similar to those so successfully employed in rearing our more valuable animals'. Yet, he observed, in the human species, where the object was of 'such consequence', the principle had been 'almost entirely overlooked'.[66]

Then in 1859 Dr Daniel Harrison Jacques (1825–77), a young American author of various 'pocket manuals' on decorum, conversation and rural crafts (including the breeding of cattle, horses and sheep), published *Hints Towards Physical Perfection*. It is a pretty good review of the mid-century angle on human improvement from a popular science position. Jacques's argument was that, through selective breeding, humans had successfully modelled the horse to plough and carry and dogs, sheep and pigeons for various purposes – both practical and aesthetic. 'Yet,' Jacques pointed out, 'we are reluctant to apply similar methods to ourselves.' For some reason, he complained, this act 'is deemed too far beyond the reach of human science and skill to be seriously posed'. Yet the examples of the ancient Greeks, as well as of modern Turks and Persians (who selected mothers for their beauty for many generations) showed it *could* be done.[67]

Modern humans are, Jacques observed, 'zealously engaged . . . in developing and molding the intellect and the affections' – so why not do the same for the physical frame? Only through 'a healthy and shapely body', he explained, could 'a sound and harmoniously developed mind' operate 'with perfect freedom and efficiency'.

His epitomes of female beauty were classical Greeks such as Diana and Helen of Troy, or 'La belle Anglaise', represented by a woman such as Lady Mary Wortley Montague. Poets such as Byron, Shelley and Keats were his ideals of modern manhood. Yet modern men and women were 'weak', 'diseased' and, 'if not absolutely ugly', then at the least 'far below our ideal standards of beauty'.[68] Jacques saw evidence for this everywhere: the 'multitude of puny and deformed children', the records of infant mortality, and the numbers crowding 'our asylums for the blind, the deaf and dumb, and the insane'. And there was the 'almost universal ill health which sustains such an army of physicians, and renders so many hospitals and water-cure resorts necessary', not to mention the 'general lack of physical vigor in both sexes, and especially in women', together with 'the scarcity of even tolerably beautiful forms and faces'. These were all pressing social and medical concerns, being articulated widely in Europe and America. All added up to prove the 'importance and urgency' of Jacques's project for physical perfection. Given this combination of facts, he wrote:

> It is certainly time for us to ask these questions and to set ourselves earnestly about the practical solution of the problems they involve. The results of our experiments upon the lower animals, and upon the products of the vegetable kingdom, point out the path to be pursued.

Progress towards physical human perfection would be gradual, 'but each step in the right direction secures an obvious and permanent benefit'.[69]

Such progress, Jacques explained, included the prolongation of life. As he observed in a chapter titled 'The secret of longevity', a quarter of all children born die before their seventh year, and half before their seventeenth. Scarce six people in a hundred lived to sixty-five, and hardly more than one in ten thousand reached their centenary: 'What an appalling picture of life – or of death, rather.' Could it be possible, Jacques asked, 'that this premature mortality is natural and inevitable?' 'Is man, as a physical being,' he wondered, 'essentially a failure?'[70]

His answer was no: 'anatomy and physiology have failed to discover anything in the structure or functions of the human system that necessarily leads to these sad results'. In fact, we are *all* born with 'the basis of a long life'. The one in ten thousand who lives a hundred years ('to say nothing', he added, 'of the still longer lives of which history and observation have taken note') is still only a human. He or she possesses 'no organs, faculties, or powers which are not the birthright of the race'. Such long lifespans are there for *all* to enjoy.[71] Writing before Thoms's damning critique, Jacques noted some recent cases of people living to 150 years and upward, including Petrarch Czarten and Louisa Truxo, 'a South American negress' who had died aged 175. And then, of course, there were those two redoubtable and ever-cited English examples, Jenkins and Parr.[72] Jacques reckoned that Buffon was wrong, and that humans lived ten times their 'growth' period of twenty years: two hundred years was perhaps the 'normal duration of life' for all people.[73]

So what exactly goes wrong? Jacques reckoned that problems started even before we were born. Through what he called the 'transgressions' of the parents, multitudes of children 'are brought into the world with the seeds of disease and dissolution already implanted in their bodies'. Others are born 'so weak or imperfect' that they 'die almost as soon as they have begun to live'. Jacques was hardly the first to warn his readers of the dangers of inherited disorders: a long tradition linked sex with disease and physical degeneration. In 1674, Sir John Pettus had observed that, by eating the forbidden fruit, Adam and Eve had discovered carnality, 'and Man hath the reward of his Libidinous disobedience, his body being so full of Disease and Infirmities, that the means of propagation seems to beget more Diseases than Children'.[74] Pettus's contemporary, the London doctor Edward Maynwaringe, suggested that one reason why the length of human lives was so much more uncertain than for other animals and plants was that we inherit the diseases of our parents, to which we then add our own; succeeding generations hence become ever 'more *degenerated*, *infirm*, *diseased*, and consequently of *shorter* duration that the former'.[75] Furthermore, 'every *generation*, have *worsted* themselves by a *degenerate* condition of life, *unsutable* [sic] to the *institutions* of Nature'. For these two reasons, Maynwaringe believed that 'we cannot but expect (unless by great *reformation* of the injurious *customs* and *vices* of these latter Ages) but that we and our *posterity* shall *degenerate* yet still into a *worse* & *sooner fading* state of life'.[76]

Nor was Jacques the first to tell his readers that they had no right to usher 'physically or mentally deformed children into a life of bodily pain or mental suffering'.[77] Here, no doubt, his reading of Hufeland influenced his thinking. In 1796 Hufeland had noted the importance that parentage plays in determining the ultimate length of our lives: irreparable damage could be done even at the moment of conception. He had himself experience of children who, 'begotten' in a 'moment of intoxication, remained stupid and idiots during their whole life'. Furthermore, Hufeland held that, whilst the father was 'without doubt' the original source of the future life, the mother was 'the soil' from which the seed derived its 'juices'. The 'future constitution' of the child thus depended upon the character of the mother, of whom the foetus was 'so long a part, and of whose flesh and blood it is actually composed'. Hufeland believed that a 'weakly father' could produce 'a robust child' if the mother was of a 'sound and vigorous body'; but even the strongest man would never produce a 'lively, healthy child' if the mother was 'weak and sickly'.[78]

This was where modern civilization had gone desperately wrong – and why no one actually lived the two hundred years that Hufeland had also calculated as the human limit. Whilst ancient nations made a pregnant woman 'a person sacred and secure from injury', the modern age encouraged women to be of a nervous, delicate constitution: 'Through that unnatural sensibility which is now so peculiar to a great part of our women,' Hufeland suggested, 'they have become far more susceptible of a thousand prejudicial effects, a multitude of passions.'

The 'fruit' that grows in the mother's womb 'suffers by every mental affection, every alarm, every cause of disease, and even by the most trifling accident'. It was therefore impossible for any child to 'acquire that degree of perfection and strength' to which it was originally destined.[79]

In Hufeland's blunt opinion, 'highly weak-nerved and sensible people ought never to marry . . . out of compassion for the miserable race of which they would be authors'. Parents should avoid educating their children too sensitively, and every man should choose a wife whose nervous system was not too 'irritable'. Women should maintain a good moral and physical regimen, and men should show appropriate respect to a pregnant woman, treating her 'with every care, tenderness, and attention'.[80]

Also influential on Jacques's thinking was Alexander Walker, a former lecturer in anatomy and physiology at the University of Edinburgh who, in 1838, published the surprisingly popular and marvellously titled *Intermarriage: Or the Mode in Which, and the Causes Why, Beauty, Health and Intellect, Result from Certain Unions, and Deformity, Disease and Insanity, from Others.*[81] According to Walker, whose book went through six editions in the USA within eighteen months, it was health, physique and the mental state of the parents that determined improvement in the human species. 'Monstrosities and diseases capable of being transmitted by generation', he advised, should be regarded 'as so many physical causes of divorce. By this means, not only sterility and deformities, but degeneration of the species would be avoided.'[82]

Only two years before Jacques's book appeared, the French doctor Bénédict Augustin Morel had published his influential *Treatise on the Physical, Intellectual and Moral Degeneration of the Human Species.* Degeneracy lurked dangerously in the modern body, Morel warned, sometimes hidden, waiting to be passed on to one's children.[83] Something had to be done. 'The means of perfecting your offspring are in your own hands,' Jacques told his readers in 1859, 'and you are responsible for their use. No child should be the offspring of weakness, or apathy, or indifference . . . much less of organic disorder, perverted passions, or brutal lusts; but of health, activity, thoughtfulness, earnestness, sincerity, purity, sweetness, harmony, and beauty.'[84] The perils of modern life were thus principally 'the creation of our own ignorance or folly'. The conditions on which longevity depended were, he advised, sevenfold:

1   A sound physical constitution (which depends upon our parents).
2   Judicious physical education (he suggested regular use of climbing ropes, parallel bars and other gymnasium exercises).
3   Simple, wholesome diet.
4   Pleasurable exercise in the open air (skating, for example).
5   Immunity from 'harrassing cares and anxieties, excesses of every kind, and all unhealthful conditions'.
6   'Constant moderate activity of body and mind.'
7   'Happiness.'[85]

Moderation in everything was the golden rule. 'If you would live long,' Jacques counselled, 'you must not live fast.' But some activity was essential. Benjamin Franklin's idea 'of suspending the processes of life entirely for a time, and afterward restoring animation', Jacques advised, 'will probably never be found practicable with the human being'.[86]

§

In the same year Daniel Jacques's *Hints Towards Physical Perfection* was published, another, far more important book appeared. Its title was *On the Origin of Species by Means of Natural Selection*; its author, of course, was the English naturalist Charles Darwin (1809–82). Darwin was educated first at Edinburgh University, where he studied medicine, and then at Cambridge, where he switched to divinity; *The Origin of Species* was the culmination of years of study. The most important part was conducted during his voyage around the world on board HMS *Beagle*, between 1831 and 1836, as companion for the ship's captain, Robert FitzRoy (1805–65).

Educated at Harrow and the Royal Naval College in Portsmouth, FitzRoy was an intelligent, accomplished seaman and surveyor; he was also deeply religious, holding rigidly to traditional ideas of natural history. Long before Darwin's famous work appeared, FitzRoy published his own account of the *Beagle*'s voyage. There he explicitly defended the authenticity of the biblical account of the Flood against the thesis recently advanced by Charles Lyell. In *Principles of Geology* (1830–3) – a book hugely influential on Darwin's thought – Lyell showed how the earth had been moulded by vast periods of slow, gradual change – a deep abyss of time almost incomprehensible to the human mind. But, for FitzRoy, Darwin's discovery, in the Andes, of petrified trees 6,000 or 7,000 feet above sea level was clear evidence of the Flood. As well as assuring readers of his voluminous report that the Deluge was a real event, he took the opportunity of pointing out that, as all believers knew, 'the life of man was very much longer than it now is'. This, for FitzRoy, was 'a singular fact, which seems to indicate some difference in atmosphere, or food, or in some other physical influence'.

Interestingly, given the evolutionary theory that his shipboard companion would gradually develop, FitzRoy did not think it very probable 'that the constitution of man' was much different at the time of the patriarchs from what it was in the nineteenth century. The reason, he thought, was that 'we see that human peculiarities are trans-mitted from father to son'.[87] But, for Darwin, change *had* happened, and it was, in part, just by this route that it had occurred: tiny changes from parent to child, over hundreds of thousands of years, had eventually resulted in substantial differences. We *weren't* like our ancestors at all. Not surprisingly, the philosophical differences between Darwin and FitzRoy became increasingly pronounced as their voyage continued, and they did not end it as friends.

His five-year circumnavigation of the world gave Darwin an incredible opportunity to study the flora and fauna of South America, Australia and the islands of the South Pacific. Then, for two decades after his return to England, he worked on his theory, making provi-sional drafts, testing his hypotheses, gathering more evidence. Lyell's *Principles of Geology*, which had helped to undermine FitzRoy's biblical explanation of the nature of the earth, was an important intellectual influence. Darwin met Lyell in 1836 and they became close friends, Lyell's conjectures suggesting to Darwin how long, long periods of gradual change could eventually effect enormous transformations – in the way a small stream of water can gouge out a deep river valley. During the *Beagle's* stopover at the Galapagos archipelago in the eastern Pacific, Darwin was intrigued to find that many animals and plants on the various islands were quite similar to one another, but 'specifically different'. He speculated that 'the inhabitants of the several islands had descended from each other, undergoing modification in the course of their descent'.[88]

The effective cause of this modification remained, however, 'an inexplicable problem' to him. It was not until his return to England – when, in October 1838, he picked up a copy of Thomas Malthus's *Essay on Population* – that the answer came to him. Then, as Darwin related in his autobiography, reading Malthus's theory of over-population, 'it at once struck me that under these circumstances favourable variations would tend to be preserved, and unfavourable ones to be destroyed. The result of this would be the formation of new species. Here then I had at last got a theory by which to work.'[89]

Charles Darwin's discovery was not evolution: as a theory, that had been around for some time. His grandfather, Erasmus, had been one proponent. The French biologist Jean Baptiste Lamarck (1744–1829) – most famously – had been another. Charles Darwin's discovery was its means and mechanism – *how* evolution happens.[90] By the dual processes of natural and sexual selection (survival of the fittest), Darwin was able to explain how species had changed over time.

But what impact did Darwin's work have on theories of ageing? In 1857, Darwin told the young natural historian Alfred Russel Wallace (1823–1913), who was close to arriving at a similar theory, that he would 'avoid the whole subject' of discussing the human species and evolution in any forthcoming book. For, though he admitted it was 'the highest and most interesting problem for the naturalist', it was, he explained, too surrounded with 'prejudices'.[91] Darwin would, however, tackle the tricky subject of human evolution (as also that of sexual selection and evolution) at some length in his 1877 book, *The Descent of Man*. There he observed that, among 'savage' tribes, 'the weak in body or mind are soon eliminated', whereas civilized people 'do our utmost to check the process of elimination'. This included the provision for such things as asylums, poor laws and medicine. Vaccination, as Darwin noted, had probably saved thousands of people

who from a weak constitution would formerly have succumbed to small-pox. Thus the weak members of civilized societies propagate their kind. No one who has attended to the breeding of domestic animals will doubt that this must be highly injurious to the race of man. It is surprising how soon a want of care, or care wrongly directed, leads to the degeneration of a domestic race; but excepting in the case of man himself, hardly any one is so ignorant as to allow his worst animals to breed.[92]

The 'unfit' thus survived, where in nature they would have perished. This suggested that progressive evolution by natural selection was a weak or even non-existent force among modern humans.

Civilization actually appeared to be working as a 'downward tendency' in human evolution: the weak survived and produced weak offspring. Certainly there were checks on this process: civilized men and women, for example, were by and large 'physically stronger than savages'. They were also more intellectually capable. But, as Darwin explained, his supporters, William Rathbone Greg (author, in 1872, of *The Enigmas of Life*) and Francis Galton, had argued: 'the reckless, degraded, and often vicious members of society, tend to increase at a quicker rate than the provident and generally virtuous members'. In *The Descent of Man*, Darwin quoted Greg, who had pointed out that the 'careless, squalid, unaspiring Irishman', who 'multiplies like rabbits', will outbreed 'the frugal, foreseeing, self-respecting, ambitious Scot', who 'passes his best years in struggle and in celibacy, marries late, and leaves few behind him'.[93] Left unchecked, and given enough time, the human race in such a state of civilization would, ultimately, degenerate. Man would get weaker, not stronger. This was not the way to perfection, or to the breeding of long life.

Darwin concluded *The Descent of Man* with the observation – which, as we have seen, he was not the first to make – that, whilst 'Man scans with scrupulous care the character and pedigree of his horses, cattle, and dogs' before he lets them mate, when it comes to his own marriage 'he rarely, or never, takes any such care . . . . Yet he might by selection do something not only for the bodily constitution and frame of his offspring, but for their intellectual and moral qualities.' Darwin, like Alexander and Jacques, thus advised that both sexes 'ought to refrain from marriage if they are in any marked degree inferior in body or mind'. He acknowledged, however, that 'such hopes are Utopian and will never be even partially realized until the laws of inheritance are thoroughly known'.[94]

On the subject of evolution and longevity, Darwin was silent. The human being, he wrote, had risen 'to the very summit of the organic scale', and this fact 'may give . . . hope for a still higher destiny in the distant future'.[95] What that future was, he did not say. In the conclusion to *The Descent of Man*, however, Darwin quashed any suggestion that humans were divinely created – we are not descendants of a perfectly formed original couple who had walked almost immortal in the Garden of Eden – and established on scientific grounds the modern consensus on human origins. As Darwin baldly put it, by considering the human embryological structure, we 'learn that man is descended from a hairy, tailed quadruped, probably arboreal in its habits, and an inhabitant of the Old World'. This quadrumanous primate was in turn descended 'through a long line of diversified forms, from some amphibian-like creature, and this again from some fish-like animal'. And, before this, it – we – were descended from some animal not unlike 'the larvæ of the existing marine Ascidians'.[96] They had not appeared fully formed, created by some divine architect; there were no seven days of creation, no Garden of Eden, no Adam and Eve, no Fall. It was not an attractive origin; but it was nearer the truth.

Darwin fully acknowledged that his theories would be 'highly distasteful to many' and 'denounced by some as highly irreligious'.[97] It is said that, at a debate over Darwin's theory, Captain FitzRoy had stood up with a copy of the Bible in his hand and shouted, 'The book! The book!' It has also been suggested that, when FitzRoy slit his throat with a razor in 1865, the depression that provoked his suicide was in part caused by regret over the role he had inadvertently played in bringing Darwin's theory into the world.

In his autobiography, Darwin wondered where evolution by natural selection – and other recent discoveries in science – left man and his famous soul. 'With respect to immortality', he wrote,

nothing shows me [so clearly] how strong and almost instinctive a belief it is, as the consideration of the view now held by most physicists, namely, that the sun with all the planets will in time grow too cold for life, unless indeed some great body dashes into the sun, and this gives it fresh life. Believing as I do that man in the distant future will be a far more perfect creature than he now is, it is an intolerable thought that he and all other sentient beings are doomed to complete annihilation after such long-continued slow progress.[98]

What, in such circumstances, was the purpose and meaning of human life? Darwin had started out as a young man intent on a career in the Church: he ended it unable to join his family at Sunday service, brooding instead in the graveyard. Yet he held on to some optimism – that phrase, redolent of Franklin, Condorcet and Jefferies, 'that man in the distant future will be a far more perfect creature than he now is'. What exactly did that mean?

§

Though Darwin was silent on the precise future state of humanity, some of his followers broached the subject directly. The Oxford University graduate Edwin Ray Lankester (1847–1929), was a bright and ambitious young man and a keen Darwinian who went on to hold chairs in zoology and comparative anatomy at the Universities of London and Oxford. In 1870 he published his prize-winning essay, *On Comparative Longevity in Man and the Lower Animals.* In his preface, Lankester noted that this was 'a subject of great popularity' and expressed the hope that his book would be of some interest to the general public. Whilst observing that he was not concerned with 'the possible means of prolonging life', he did note Bacon's *History of Life and Death* with approval. Though it contained some 'strange fancies', he considered it 'a most admirable enquiry into the causes of longevity'. He noted, too, that Hufeland had modelled his *Art of Prolonging the Life of Man* on Bacon's essay, and that though Hufeland 'did not accomplish his task, which indeed he could not hope to do', he had 'shewn an excellent path, which it remains for others to improve and extend'. Other works cited included what Lankester called Sir Henry Holland's 'able essay' on human longevity – the very one which was so ably dismissed by Thoms. If Lankester took Holland seriously, then there were going to be some serious factual limitations to his conclusions – as we shall see.[99]

Lankester posed that recurrent question: what is the limit of the length of human life, and could death be connected to whatever it is that stops us from continually growing? He based his answer to the second part of this question on the theory expressed in the philosopher Herbert Spencer's *Principles of Biology* (1864–7). The ageing process, Lankester explained, paraphrasing Spencer, is all to do with protoplasm. This was 'the physiological basis of life' and the 'germinal matter'. As the latter accumulates, an organism increases its power; as it declines, the amount of life diminishes. This so-called 'germinal matter' forms tissue, processes repair, manufactures secretions, effects force. It is more abundant in youth, increasing from conception; but it does not increase as fast as the whole organism, and thus it 'is always diminishing relatively to the whole'. Eventually, there is insufficient germinal matter to keep up with physical growth – and it is no longer accumulated, but destroyed: 'thus the inherent cause of death has a structural existence'. This is why trees, reptiles and fish are long-lived: because they have very little 'personal expenditure' of germinal matter. Furthermore, when equilibrium between accumulation and expenditure is reached, a 'very severe tax' on germinal matter kicks in – reproduction. 'The effect of this additional tax is to start the organism more rapidly down the incline towards the termination of the road of life . . . the earlier reproduction is commenced, and the more rapidly it is carried on, the sooner must the increase of the organism's bulk be stopped, and so waste and death ensue.'

In terms of maximum lifespan, Lankester accepted Buffon's statement that all races of men, from black to white, civilized to savage, could live at most only between ninety and one hundred years: 'Hence we do not look for *much* differences of longevity, even in different climes and different civilizations.'[100] Lankester, therefore, did not lay much emphasis on individual cases of longevity. These, he believed, were of 'little scientific value', and were to be considered as cases of what he called 'abnormal longevity'. They included a case quoted in a recent note in *The Lancet*, forwarded from the American Philosophical Society. Based on life tables, it indicated 'a possible life in Philadelphia of 114 years'.[101] He also noted the personal communication of Sir Henry Holland, who had told him of men he had met who possessed documented evidence that they were 104 and even 111 years old. Lankester thus believed it *was* possible for some people to achieve a higher longevity than a hundred years, even up to 120 years. But, as with the rare instances of human gigantism, he considered any such to be 'a monstrous and abnormal phenomenon'.

Figure 4.2 Edwin Ray Lankester: colour lithograph after Sir Leslie Matthew Ward ('Spy'). In his prize-winning essay, *On Comparative Longevity in Man and the Lower Animals* (1870), the young zoologist Lankester applied Darwinian ideas to the question of life and death. He believed that an 'almost perfect civilization' might eventually evolve in which everyone lived a hundred years or more.

Such examples of longevity were freaks of nature transcending its more universal rules.

Setting aside such anomalies, could natural selection work actually to increase longevity? Lankester felt that, given modern civilization's tendency to call upon 'increased mental expenditure', it seemed possible that 'future men' would 'rather lose than gain in longevity'. Stress and mental exertion, as we have seen, were not considered conducive to long life. But he was confident that the civilized human's 'structural capacity' for coping with these would 'increase simultaneously'. The modern human was living 'in the midst of a struggle', he believed, which was operating not only on communities 'but also on individuals, and by means of this struggle greater mental power is being added to the human race'. Contrary to his earlier statement, it appeared to Lankester that, though the potential longevity of humans was much the same for the various races, '[t]he highest civilization, corresponding to the highest evolution, appears to give a somewhat increased potential longevity'. The 'apparently higher longevity' of the English, as contrasted with members of other western European nations, seemed (to him) suggestive of this fact. Lankester believed that it might be due to 'a somewhat higher development' in the English character. The shorter lives of the people of the USA, however, he found puzzling. Using 'their own expressive phrase', he suggested it might have something to do with the fact that 'they are a "go-a-head" people' with 'an individual tendency to travel fast as regards age'.[102]

Echoing (probably subconsciously) the ideas of Condorcet, Lankester believed that, from this contemporary struggle, an 'almost perfect civilization' would eventually result (he put no time-scale on it), in circumstances 'where the greatest happiness for every individual must finally be attained'. Would longevity be extended in such a civilization? It did not seem improbable. At first, he predicted, everyone would 'attain eighty or a hundred years'. But he saw 'no apparent reason why longevity should not increase beyond that limit, and advance with advanced evolution, and the diminished expenditure implied in more complete adjustment'.[103] Darwin read Lankester's book and told him he was 'much interested' by it. Though Darwin acknowledged that there had to be a great deal of speculation in such a subject, 'all your views', he informed Lankester, 'are highly suggestive'.[104] Darwin cited the work approvingly in the *Descent of Man*.

The promising prospect of future evolution was counterbalanced, however, by the threat of degeneration – that looming Victorian spectre of species evolving *backwards*. The notion of degeneration permeated nineteenth-century thought. It has been described as at once 'the most frightening of prospects, as well as at times the most enthralling'.[105] In literary terms, it gave rise to such classics of gothic fiction as Robert Louis Stevenson's 1886 novella, *The Strange Case of Dr Jekyll and Mr Hyde*, and Bram Stoker's 1897 classic blood-sucking tale of living-dead vampires, *Dracula*. Edwin Lankester laid out a more scholarly analysis in a short book, *Degeneration: A Chapter in Darwinism*, published in 1880. There he noted that progress of the human form was neither necessary nor inevitable. Compared with the ancient Greeks, humans did not appear to have improved either in bodily structure or mental capacities over the preceding two and a half millennia. Perhaps even now, he wrote, the English were degenerating 'into a contented life of material enjoyment accompanied by ignorance and superstition'.[106]

In 1883, Darwin's cousin, Francis Galton, coined a term for the means by which human degeneration could be avoided and progress achieved: he called it eugenics. Deliberate government policies would help to manipulate the race's biological constitution.[107] The criminal classes, the congenitally diseased and the insane would be prevented from reproducing, whilst the 'fitter' would be encouraged to have more children.

In an extraordinary book, *Degeneration*, published in the 1890s, the German sexologist Max Nordau described decadence and the *fin-de-siècle* mood as 'the impotent despair of a sick man, who feels himself dying by inches in the midst of an eternally living nature blooming insolently forever'. To Nordau, premature death, decay and insanity appeared to characterize modern existence: 'We stand now in the midst of a severe mental epidemic,' he declared, 'a sort of black death of degeneration and hysteria, and it is natural that we should ask anxiously on all sides: "What is to come next?"'[108]

Civilization, it appeared to some critics, was turning backwards before their eyes. The nineteenth century ended on a downbeat note.

CHAPTER 5

✷

A Brave New World

Biology has not yet been able to decide whether death is the inevitable fate of every living being or whether it is only a regular but yet perhaps avoidable event in life. It is true that the statement 'All men are mortal' is paraded in text-books of logic as an example of a general proposition; but no human being really grasps it, and our unconscious has as little use now as it ever had for the idea of its own mortality.

Sigmund Freud, 'The "Uncanny"' [1919], in the *Standard Edition of the Complete Psychological Works* (1955)

France, a little after dawn, 1 July 1916 Above the monotonous rumble of heavy artillery that has continued day and night for weeks, there's an ear-splitting roar. The explosion rips the ground. It is followed minutes later by more colossal blasts. The whole world feels like it's moving. Vast plumes of smoke, chalk dust and debris are thrown high into the sky. At a stroke, hundreds of lives are annihilated.

Two minutes more, and thousands of British and imperial soldiers rise from the earth and press forward their attack: the massive mines are blown, the artillery barrage rolls forward, and the 'tac-tac-tac' of German machine-guns joins the unholy row. Men fall as gunfire sweeps their slow-moving ranks. Kilted Highlanders are caught on barbed wire hedges, their bodies hanging like dead crows in the sun, punctured repeatedly by bullets; Newfoundlanders and Tyneside Irish crossing open ground are mown down before they reach the forward trenches; death falls unequally in No Man's Land. 'It was about this time', recalls one fortunate survivor, 'that my feeling of confidence was replaced by an acceptance of the fact that I had been sent here to die.'[1]

It is mass slaughter on an industrial scale, social Darwinism in action, in which only the fittest race (some suppose) will emerge victorious. By the end of the first day of the Battle of the Somme, over 20,000 British and Commonwealth soldiers are dead, killed in a war that will on average slaughter 5,000 combatants *every single day* for over four years.

Thirty miles to the west, in Paris, an elderly Russian biologist, a gaunt, mountain of a man with dishevelled hair and a long white beard, lies dying. Born on 16 May 1845 in what is now eastern Ukraine, by the age of six Ilya Ilyich Metchnikoff was lecturing his family and friends on the local flora and fauna. At school he became a microscopist and an atheist, coming firmly to hold the idea that the progress of civilization depended upon the advancement of science, not upon theology. He published his first scientific paper aged eighteen; at twenty-two he was appointed Professor of Zoology at the new University at Odessa. In Russia, Germany and Italy, in a scientific atmosphere dominated by the works of Charles Darwin, he studied comparative embryology and evolution. Depressive and at times suicidal, he determined in his twenties never to have children – but, for companionship's sake, he would marry twice.

Significant advances had been made in medicine through the nineteenth century; Metchnikoff's life's work would fall within their ambit. In the late 1830s, the German biologist Matthias Schleiden (1804–81) and the German physiologist Theodor Schwann (1810–82) identified cells as the fundamental units of life. In due course, the pathologist Rudolf Virchow (1821–1902) used improved microscopes to examine the body at its most minute level. He too focused on cells, rather than organs, as the source of disease. His work saw a sea change in the understanding of disease processes.[2] The cell became a locus of research on ageing: in 1891, the German biologist August Weismann (1834–1914) suggested that 'death takes place because a worn-out tissue cannot forever renew itself, and because a capacity for increase by means of cell division is not everlasting but finite'. Ageing and death were evolutionary processes, nature's way of eliminating the old, who competed with the young for scarce resources.[3]

The German chemist Justus von Liebig (1803–73) and the French microbiologist Louis Pasteur (1822–95) made further headway in understanding the operation of disease. Since ancient Greece, the patient (understandably) had dominated medical treatment in the West: under the Romans, Galenic medicine focused its attention on the four humours, with health or sickness seen to depend upon each individual's circumstances. Now, doctors and chemists increasingly turned their attention on *diseases*. It is the disease, not the patient, that is to become the focus of study: know the disease, and you can implement the cure – or even prevent the disease.

In the 1790s Edward Jenner had discovered what would be the world's first vaccination – using cowpox to confer immunity to smallpox; this was quickly heralded as a major advance in medical science. How it worked, however, was a mystery. But, through diligent experimental work, Pasteur helped rapidly advance the understanding of the role bacteria play in the development of many diseases. He showed that fermentation and putrefaction were biological processes dependent on the action of specific living micro-organisms. Realizing that bacteria invading the system and quickly multiplying within the body cause diseases such as rabies and anthrax, he helped develop the germ theory of disease. In 1882, a former pupil of Rudolf Virchow, the physician Robert Koch (1843–1910), announced that tuberculosis, one of the most dreaded infectious diseases, was caused by a bacterium that he had isolated and identified. Micro-organisms were soon recognized as the origin of a range of killer diseases: smallpox, cholera, diphtheria, typhoid fever. Then Pasteur discovers a vaccine for rabies, whilst one of Koch's former assistants, Emil von Behring (1854–1917), isolates the diphtheria bacillus and produces one of the first antitoxins. Between 1872 and 1903, the causative organisms of almost every common disease are identified; this is truly a revolution in the biological sciences.[4] The sciences of bacteriology, immunology and biochemistry have been born.[5]

In 1888, Metchnikoff moves to Paris, where he becomes head of the department of microbiological morphology at the newly established Institut Pasteur. There he studies the mechanisms of immunology and vaccination. By the late 1890s, his scientific interests in immunology and infectious diseases are increasingly directing his attention to the causes of ageing. In 1901, Metchnikoff publicly airs his ideas for the

Figure 5.1 'Professor Metchnikoff': caricature by B. Moloch from *Chanteclair*, 1908, no 4. By 1908, Ilya Ilyich Metchnikoff's name was synonymous with yoghurt and schemes for the prolongation of life beyond a hundred years. In this French cartoon, the Nobel-prize winning 'manufacturer of centenarians' is seen feeding industrial quantities of curdled milk to geriatric patients, who suck at the tubes like infants.

first time at a lecture in Manchester. Two books detailing his theories soon follow: *The Nature of Man* (1903) and *The Prolongation of Life: Optimistic Studies* (1907). And he coins a new word to describe this research: he calls it 'gerontology'.

The first gerontological question Metchnikoff has to answer is: What is the 'natural' maximum lifespan of the human being? In 1899 the British actuary Thomas Emley Young had published an authoritative exploration of just this question, *On Centenarians and the Duration of the Human Race*. From a base of close on one million records collected by the Life Assurance and Annuity Societies of

Great Britain, as well as from records from the National Debt Office, Young found only thirty persons who had lived for a century or more. The oldest he felt able to authenticate was a woman who had died just short of her 111th birthday.[6] Unfortunately, Metchnikoff knows nothing of Young's book, and conducts his own research. A journalist introduces him to a Parisian woman, Mme Robineau, who (it is said) is 107. Though her physical condition shows 'extreme decay' – the skin of her hands are so transparent Metchnikoff can see through it to the bones, blood vessels and tendons beneath – her mind is delicate and refined, her conversation intelligent, connected, logical.[7]

Unaware of Thoms and Young's warning not to rely on hearsay, Metchnikoff does not check the authenticity of Mme Robineau's great age. With this remarkable example of longevity, he rejects suggestions that the natural limit of human life is only seventy or seventy-five years. He claims that centenarians are 'really not rare' at all: in France alone, he reckons, 'nearly 150 people die every year, after having reached the age of 100 or more'.[8] Though Methuselah's 900-plus years is clearly 'a mistake in calculation', there are other examples. He notes the cases of Drakenberg, who died aged 146, and the 'well-known instance of Thomas Parr'. This, he points out, rests 'on good authority', since William Harvey had autopsied the body. 'It appears,' Metchnikoff concludes, 'that human beings may reach the age of 150.' But, he notes, such cases are 'extremely rare' – none being known 'from the records of the last two centuries'.[9]

Basing his argument partly on what has already been proven spurious evidence, it appears to Metchnikoff that few humans die 'natural' deaths. Senescence and death by seventy or eighty, he suggests, is a pathological phenomenon, in disharmony with nature. Death at this age is not 'instinctive'. This is a novel theory, and Metchnikoff puts great emphasis on it. After a hard day's work, he explains, we desire rest; when we are hungry we want food, and, once full, we do not wish to eat more. Yet we have no similar wish to die – at least, not in youth or middle age. He is particularly moved by an observation made by the writer Anthelme Brillat-Savarin, who had visited his ninety-three-year-old great aunt on her deathbed. The old lady told her nephew: 'If you ever get to my age you will see that death becomes as necessary as sleep.'[10]

For Metchnikoff, that most of us die *prematurely* is confirmed by this 'fact that old people who have reached an exceptionally advanced age are often satiated with life and feel the *need* of death'. It is for this reason that 'we have a right to suppose that, when the limit of life had been extended, owing to scientific progress, the instinct of death will have time to develop normally and will take the place of the fear which death provokes at the present day'.[11] Then, at 150 or so, '[m]an, having been through his normal vital cycle, will sink, peacefully and without fear, into eternal sleep'.[12]

In support of his theory Metchnikoff notes that, as he has grown older, his own lust for life has gradually diminished. He feels sure he will eventually accept his own demise with equanimity. The Austrian neurologist Sigmund Freud, influenced by Weismann's work and the Great War, is developing a similar theory of the 'death instinct'. 'What lives', Freud will write in *Beyond the Pleasure Principle* (1919–20), 'wants to die again. Originating in dust, it wants to be dust again.'[13]

Even so, Metchnikoff notes that death is not necessarily inevitable: many micro-organisms, which reproduce by splitting into two, are effectively immortal. And some higher plants, such as the dragon tree in the Canary Islands, the baobab trees of Cape Verde, and the *Sequoia gigantea* of California, seemingly die only by accident. There is 'nothing', he concludes, 'to be found in the nature of their organisation which would seem to indicate that death is the inevitable or even probable result of their constitutions'.[14]

If death is not essential, what explains it? Why die at all? He rejects Weismann's suggestion that ageing is caused by the eventual inability of cells to reproduce: even when we are old, he points out, we continue to grow nails and hair; indeed, the latter can grow even more vigorously, as in old women who grow facial hair.[15] Understandably, given the nature of his own researches and those of the Institut Pasteur, Metchnikoff wonders whether bacteria somehow cause senescence. Unicellular beings, he has found, are generally immune to infectious diseases: with bodies almost entirely made up of digestive protoplasm, they either destroy the microbes they absorb, or immediately expel them. The study of immunity, he suggests, is, ultimately, all about digestion. He develops the thesis that the cause of ageing is chronic poisoning by types of cells called phagocytes, which, as his wife later explains, 'absorb all the enfeebled cells in the organism'.[16]

To try and prove this theory, he investigates one of the clearest signs of human ageing: greying hair. Under his microscope, he finds the nerve cells in the hair follicle appear to be under attack by what he calls neurophages, which absorb their contents and bring about 'more or less complete atrophy'. A colleague, the Romanian neurologist George Marinesco, doubtful of Metchnikoff's theory, sends him samples from the brains of two human centenarians – one (he says) had died aged 117. After conducting a careful examination, Metchnikoff concludes that phagocytes surround many of the nerve cells in these elderly brains, which were in the process of being destroyed by them. Further proof, he insists, of his hypothesis.[17]

Many of these destructive microbes, he suggests, originate in the gut. The evidence seems clear: it is, above all, what he calls 'intestinal putrefaction' that shortens and ultimately terminates life. Ageing and death are caused by the invasion of the body's cells by billions of microscopic organisms. He rejects the theory that these bacteria play some unknown but positive role in digestion; he raises animals in his laboratory in sterile conditions to 'prove' that intestinal flora are wholly unnecessary for normal life functions; he examines the intestinal effects of cholera and typhoid on monkeys to show their deleterious consequences.[18]

These harmful bacteria, Metchnikoff argues, are particularly prevalent in humans. Our extensive large intestine, though useful early in our evolution, is now redundant – and deadly. Indeed, from his studies in comparative anatomy and longevity, Metchnikoff concludes that the longer an organism's large intestine, the shorter its life. Natural selection, he argues, had not created a perfected animal. We are, in fact, what he considers 'an ape's monster': we possess oversized brain and hands (good things), but also a large intestine (a bad thing) – a throwback to an earlier time in our evolution, when stopping regularly to defecate when on the run from predators would have been dangerous.

The large intestine, therefore, simply functions as a 'reservoir' for 'waste'. Faecal matter could be stored until it could be expelled at a time and place of safety.[19] The micro-organisms that thrive off this putrefying material – like cholera bacilli multiplying in soiled water – slowly attack us from within, poisoning us by degrees. Diseases such as syphilis and pneumonia likewise weaken the body, gradually grinding down its capacity to resist infection and decay. As he writes in 1910, 'the human being, in that state in which he has appeared on earth, is an abnormal, sick creature'.[20] Disease is in our essence.

Having located the problem, what is the cure? What is needed, Metchnikoff explains, is a 'remedy' that can strengthen the 'higher elements' of our physiques, making them less ready prey to the microbes that 'devour' us.[21] His first answer to solving the problem of premature death by auto-intoxication is sour milk. Intestinal microbes need an alkaline context to survive, but putrefaction in sour milk is prevented by the latter's natural acidic fermentation. Sour milk, with its acid-producing microbes, would prevent noxious microbes from breeding in the gut, reducing their negative effects. His hypothesis seems to be confirmed by the fact that some people (such as the Bulgarians) who feed 'almost exclusively on cured milk live a very long time'.[22] He immediately introduces sour milk into his diet. He recommends his friends to do likewise, and promotes its use in his books.

But sour milk is no panacea. Metchnikoff is well aware that, on its own, it can be no antidote to a bad diet. So sour milk is only the *first step* in a process that will see the artificial transformation of wild intestinal flora into cultivated, beneficial ones. It will form part of a carefully controlled regimen – something Metchnikoff also explores. Working with collaborators at the Institut Pasteur, he uses rats to study the differing physical effects of carnivorous, vegetarian and omnivorous diets – finding the latter to be best for health. Any discovery that seems relevant to prolonging human life is adopted into his daily practice.

The Darwinian scientist and writer on longevity Earl Ray Lankester had become close friends with Metchnikoff in the 1890s. He notes that the Russian biologist never smokes or drinks, is fastidious about the potential contamination of uncooked food by bacteria and parasites, and never eats raw vegetables or unpeeled fruit. 'This was not an excess of caution,' Lankester explains, 'but resulted from his characteristic determination to carry out in practice the directions given by definite scientific knowledge, and to make the attempt to lead so far as possible a life free from disease.' Even when they lunch together at the best cafés in Paris, Metchnikoff eats 'very simple food'.[23] Lankester for a while takes to drinking soured milk – as do other of the Russian's friends.

'Orthobiosis' is Metchnikoff's word for this programme that will prolong life to a 'natural death'. It will require, he acknowledges, numerous modifications of our existence, including 'an active, healthy, and sober life, devoid of luxury and excess'. But he believes that, as 'the struggle against diseases, the prolongation of human life, and the suppression of war make progress', humans will feel less need to reproduce. Reducing overpopulation will be 'one of the chief means of diminishing the most brutal forms of the struggle for existence, and of increasing moral conduct amongst mankind'.[24]

Long troubled with a bad heart, Metchnikoff realizes that what is likely to be his own, comparatively early, death may undermine his theory. In an autobiographical statement written after a heart attack in 1913, he asks 'those who think that, according to my principles, I should have lived a hundred years' to 'forgive' him his 'premature end'. There were, he explains, 'extenuating circumstances': the intensity with which he has pursued his scientific studies from a very young age; his excitable temperament; his nervous disposition; and the lateness with which he had started the dietary programme that one day would see all people enjoy natural lifespans well beyond a century. He notes, too, that he has already lived much longer than any other member of his family.[25]

In 1906 he receives the prestigious Copley Medal from the Royal Society of London for his work in zoology and pathology. The Nobel Prize for medicine and physiology follows in 1908, in recognition of discoveries in immunology. In 1911, the *Encyclopaedia Britannica* records that Metchnikoff's suggestion 'that duration of life may be prolonged by measures directed against intestinal putrefaction' appeared 'rapidly to be gaining ground' – though it adds that it is 'yet far from being accepted'.[26] But, if true, it has exciting implications. Dr Isaac Max Rubinow, the Russian-born pioneer of the USA social security movement, writes in 1913 how '[t]he great discoveries of Metchnikoff have given the human race a new hope, a new conception of what the normal span of life ought to be'.[27] In 1914, *The Times* reports that Metchnikoff's 'theories on longevity' are 'well known', and notes that the biologist is 'notoriously vigorous for his years and seems to find no need for holidays'.[28] His recommendation of sour milk gains public interest: a French manufacturer begins production on a large scale under the approved by-line 'Sole provider of Professor Metchnikoff'.[29]

In 1915, Metchnikoff celebrates his seventieth birthday. Though an atheist, he is pleased to have reached the three score years and ten of the psalm. But in the future, when macrobiotics has been perfected, he believes that 'the normal limit of life will be put much further back and may extend to twice my 70 years'. Science, he confidently believes, will one day 'solve all the principal problems of Life and Death'.[30]

But the Great War grieves Metchnikoff profoundly. Every newspaper he reads saps his will to go on. The Institut Pasteur, where he devoted all his efforts to researching means of preserving and prolonging life, has been emptied and turned over to the service of the conflict. His young students and doctors are at the front. These 'brilliant young lives', his wife later laments, 'sacrificed, victims of those who should have directed the peoples towards peace and a rational life, and who, instead of that, threw the most precious part of humanity into the abyss of death'. War, she records, became 'a dark, sinister background' to her husband's daily life. Its casualties 'were not only those who fell on the battle-field'. They included Metchnikoff himself, 'whose whole life-effort had been directed towards the conservation of human existence and the search for rational conceptions. The contrast between his aspirations and the cruel reality had been to him a blow which his sensitive and suffering heart was not fit to bear.'[31]

Yet even as he is dying, Elie Metchnikoff is working on one last book, a history of modern medicine. Only the preface will ever be written. There he records his hope that this 'insane war' with its 'unexampled butchery will, for a long time, do away with the desire for fighting, and that soon the need will be felt of a more rational activity'. Instead of fighting each other, men of a better future will struggle together to fight 'the innumerable microbes, visible or invisible, which threaten us on all sides and prevent us from accomplishing the normal and complete cycle of our existence'.[32]

On 15 July, the old man dies. According to his wife, the doctors who carried out the postmortem wonder how he had lived so long with such a poor heart. They put his survival down to the strict regime he had followed in the latter part of his life.[33] Metchnikoff's remains are cremated and his ashes deposited in the library of the Institut Pasteur. On the Somme, the British offensive continues bloodily, remorselessly – sucking life up in its brutal, machine-like and indifferent grasp.

§

Metchnikoff's theory excited a great deal of interest, and other scientists undertook further investigations. Alexandre Besredka was a young officer at a French military hospital when he heard the news of Metchnikoff's death; he later recalled how conversation in the mess became very lively when his colleagues started discussing the biologist's promotion of sour milk. Besredka went on to work at the Institut Pasteur, and declared in 1921 that with Metchnikoff's work a new philosophy had appeared, 'a new concept of life destined to change our most firmly anchored ideas'. Metchnikoff did what Francis Bacon had done several centuries earlier 'for the sciences in general, proclaiming the necessity of systematic studies'; he 'proposed the same reforms in philosophy, in particular, to the study of old age'.[34]

Minoru Shirota, a Japanese medical student who began his studies at Kyoto University in 1921, was also influenced by Metchnikoff. Shirota became 'convinced that healthy intestines contributed substantially to a healthy life', and studied intestinal bacteria. He eventually isolated a lactic acid bacterium that, according to the company he went on to found, 'would help to protect the gut by maintaining the natural balance of the intestinal flora'.[35] In 1935 Shirota developed a 'probiotic' fermented milk drink, Yakult, which is still widely promoted on TV.

Though probiotic yoghurts have endured as supposedly health-benefiting food, Metchnikoff's theory has not. Further research proved his theory wrong. Towards the end of the Great War, the biochemists Jacques Loeb (1859–1924) and John Northrop (1891–1987) of the Rockefeller Institute for Medical Research studied the lifespans of fruit flies raised in conditions entirely free from bacterial contamination. Compared with fruit flies bred in the laboratory of Raymond Pearl at Johns Hopkins University in Baltimore, the sterile flies actually had a *shorter* mean lifespan. By comparison with what he considered to be the much greater influence exerted by heredity, Pearl concluded in 1920 that bacteria 'play but an essentially accidental rôle in determining length of the span of life'.[36]

As the pioneering work of Loeb, Northrop and Pearl indicates, senescence and rejuvenation were subjects of profound medical interest in the first decades of the twentieth century. In the aftermath of the Great War's slaughter, there emerged an understandable obsession with prolonging and reinvigorating life. In October 1920, an editorial in the *Journal of the American Medical Association* criticized this phenomenon. 'At present,' it complained, 'there seems to be a sort of international scrambling for priority recognition in the alleged discovery of the profound secret of restoring lost youth and youthful vigor . . . One must marvel at the ease with which fragmentary data are woven into a story of technical success.'[37] Yet the editorial did not suggest that these restorative measures might not work. It only advised that more information was required before jumping to far-fetched conclusions.

The man who had set the ball rolling, even before Metchnikoff, was the Franco-American physiologist Charles-Edouard Brown-Séquard (1817–94). In 1889, after a distinguished career, including professorships at the Collège de France and Harvard, the seventy-two-year-old Brown-Séquard announced before a meeting of the Société de Biologie in Paris that he had succeeded in rejuvenating himself. His method was founded on what he called the 'well known' fact 'that seminal losses, arising from any cause, produce a mental and physical debility which is in proportion to their frequency'. (This was the medical theory behind the Victorians' castigation of masturbation.) He explained that 'well-organised men', especially those aged between twenty and thirty-five, who avoided *any* sexual activity, were 'in a state of excitement' that gave them 'a great, almost abnormal, physical and mental activity'. This led Brown-Séquard to conclude that there existed in seminal fluid substances 'which, entering the blood by resorption [sic], have a most essential use in giving strength to the nervous system and to other parts'.

Brown-Séquard spent some time seeking ways of artificially restoring this substance. Whilst living in Massachusetts in the 1870s, he had experimented with grafting parts of young guinea pigs onto dogs. Though only one case had shown favourable results, this apparent success had given him hopes that his transplantive method had validity. He now claimed before the *société* that he had had success the previous year using rabbits, and that he had recently experimented on himself.

He had, he explained, produced a rejuvenating liquid by mixing small quantities of distilled water with blood taken from the testicular veins of young and healthy dogs and guinea pigs: to this he had added semen and juices from their crushed testicles. He had filtered this mixture, and on ten occasions between 15 May and 4 June 1889 injected a total of 10 cubic centimetres subcutaneously into his arms and legs. Though there had been some 'pains and other bad effects', he asserted that both animals 'have given a liquid endowed with very great power'.

As Brown-Séquard subsequently reported in *The Lancet*, over the previous ten or twelve years his general strength had 'notably and gradually diminished'. He was often exhausted after working in his laboratory, would go to bed early, and woke up 'exceedingly tired'. But after only three injections of his rejuvenating formula 'a radical change' had taken place. He was no longer tired; he could come home and continue working after dinner, as if he was twenty years younger; he had regained strength in his limbs, and could urinate more strongly. His intellectual powers had also improved: it was 'evident' to him 'that all the functions depending on the power of action of the nervous centres, and especially of the spinal cord, were notably and rapidly improved'.

Noting the English physician Daniel Hack Tuke's recent work on the influence of the mind on the body,[38] Brown-Séquard invited other old men to carry out the same experiment. He wanted to be sure that the effects he had experienced were not dependent on 'any special idiosyncrasy or on a kind of auto-suggestion without hypnotisation, due to the conviction which I had before experimenting that I should surely obtain a great part at least of these effects'.[39] Though the effects of the injections had started to wear off after four weeks, he believed it was possible that, if continued long enough, 'some structural changes' in the body, 'not essentially allied with old age, although accompanying it, [would] disappear to such a degree as to allow tissues to recover the power they possessed at a much less advanced age'.[40] Further researches by others, he hoped, would answer these questions.

The claims of the ageing French physiologist were not universally well met. According to one German scientific journal, these 'fantastic experiments with testicular extracts must be regarded almost as senile aberrations'.[41] Dr Edward Berdoe, a licentiate of the Royal College of Physicians and an ardent opponent of vivisection, was soon circulating a letter declaring that Brown-Séquard's 'abominable proceedings' were undertaken simply in order 'to enable broken down libertines to pursue with renewed vigour the excesses of their youth'.[42]

Some of the scepticism greeting Brown-Séquard's claims derived, no doubt, from the fact that using testicular matter for sexual rejuvenation had a long tradition. As far back as the eighth century BC, the Indian Samhita of Sushruta had recommended that a man wanting to satisfy a hundred women should eat the testes of goats, either boiled in milk or cooked in butter.[43] Furthermore, physiologically speaking, we now know that Brown-Séquard's claims were simply impossible. The human body cannot accept or assimilate animal tissue in that way without the assistance of modern drugs. Any rejuvenation he experienced must have been purely psychological – a self-induced placebo effect, the danger of which he had been all too aware. Within five years of his experiments in rejuvenation, and some three years short of his eightieth birthday, Brown-Séquard was dead.

This is not to say that Brown-Séquard's theory was wholly irrational, or that it did not inspire other scientists to follow his lead. The study of hormones (a word not coined till 1905, from a Greek verb meaning 'to stir (into action)', 'excite' or 'arouse') was a new science, and Brown-Séquard one of its leading instigators. His experimental studies of the adrenal glands in the 1850s 'initiated the science of modern experimental endocrinology',[44] and in the mid-1860s he had been Harvard's first professor of the physiology and pathology of the nervous system. He had gone on to suggest that many diseases could be caused by glandular deficiencies, and it seemed logical to him – as well as to others – that extracts from the testicular and other endocrine glands could be used to cure them.[45]

Whilst charlatans hoping to mint quick money were soon making extravagant claims for an 'elixir of life' supposedly based on Brown-Séquard's technique, researchers conducted their own therapeutic trials. Physicians in Russia, Poland, Austria and Italy obtained free supplies of Brown-Séquard's animal testicular extract – and reported promising results.[46] British doctors who explored Brown-Séquard's theory included the neurosurgeon Sir Victor Horsley (1857–1916) and his student George Murray. In 1891, Murray announced that he had used hypodermic injections of extracts of thyroid tissue to cure thyroid deficiency in sheep.[47] Two years later, George Oliver, a Harrogate physician, and Edward Schäfer, professor of physiology at University College, London, showed that the adrenal glands contained a substance which produced dramatic pharmacological effects.[48]

Though the German scientist Fürbringer subsequently discredited Brown-Séquard's work, and Metchnikoff would note (with a hint of disappointment) that 'the observations of physicians, made on old men and sick persons, have not justified the hopes which were entertained on the mode of treatment',[49] Murray's success with thyroid glands and Oliver and Schäfer's with what, when isolated, became known as adrenaline, lent credibility to this whole line of research. Whilst the *British Medical Journal* remained cautious in its assessment of Brown-Séquard's work, these were real medical breakthroughs in the new field of endocrinology. The discovery of the pancreatic hormone insulin in the 1920s, with its powerful effect on diabetics, would renew excitement. As Chandak Sengoopta has observed in his book *The Most Secret Quintessence of Life* (2006), 'the concept of internal secretions pushed medical thinking in bold new directions'.[50]

§

To follow the course of endocrinology into the twentieth century, it is helpful to look at the career of the French surgeon Alexis Carrel (1873–1944). After completing medical training in France, Carrel moved to the USA, working first in Chicago and then at the Rockefeller Institute for Medical Research in New York. In 1910 he announced that he and his colleagues had succeeded in keeping cultures of animal and human tissue – including some from the thyroid and renal glands – alive *in vitro*. Removed from the body of which they had originally been a part, these cell cultures lived, fed, grew. Prior to the team's work, such isolated cells had always withered and died within about fifteen days.

In a paper published in 1912, 'On the permanent life of tissues outside the organism', Carrel explained that the purpose of these experiments had been 'to determine the conditions under which the active life of a tissue outside of the organism could be prolonged indefinitely'.[51] His success was a remarkable achievement, and it led him to declare that 'senescence and death are a contingent and not a necessary phenomenon'.[52] In that same year, Carrel established a subculture of chick embryonic heart fibroblasts which would seemingly continue to survive *in vitro* for a staggering thirty-four years – well beyond the lifespan of any normal chicken.[53] This suggested to Carrel that, in the same way that most bacteria and protozoa are considered immortal, tissue cells kept *in vitro* and treated under the correct laboratory conditions could continue growing and dividing indefinitely.[54] It appeared that ageing and death were features only of *multi*-cellular organisms. As Raymond Pearl put it in his 1922 book *The Biology of Death*, Carrel's experiment had demonstrated 'the potential immortality of somatic cells, when removed from the body to conditions which permit of their continued existence. Somatic cells have lived and are still living outside the body for a far longer time than the normal duration of life of the species from which they came.'[55]

In 1912 Carrel won the Nobel Prize for Medicine for his surgical work, including that into the field of accelerating the healing of wounds. The public response illustrates the role which the popular press increasingly played in building up hopes of medical discovery in the twentieth century. In 1913, a Philadelphia newspaper reported excitedly that, whilst Carrel 'doesn't come right out and say so, he leads us to believe that in future we will be quite exempt from all bodily injuries'.[56] In 1927, Antoni Nemilow, professor of anatomy and cell research in Leningrad, declared that Carrel's apparently immortal embryonic chicken cell culture 'proved clearly that science is stronger than death ... This growing, living tissue unconditionally contradicts, once and for all, superstition and all fairy tales of the higher force of death, and stands above human will and reason.'[57]

More recent research (which we will discuss in the next chapter) has shown that such cultures cannot survive for more than sixty to eighty days *in vitro*, and the endurance of Carrel's so-called '"immortal" strain' remains something of a mystery.[58] But the work of the Nobel prize-winning French surgeon played a hugely influential role in the development of (inaccurate) theories of cell ageing into the early 1960s. Failure to replicate Carrel's results in other laboratories was blamed on inadequate culture conditions or incompetence. For many decades it led most scientists to believe (wrongly) that senescence worked *not* at the individual cellular level, but that it was purely a macro-phenomenon.

The American humanist philosopher Corliss Lamont (1902–95) discussed these developments in his 1936 book *The Illusion of Immortality*. Though essentially a scientific rejection of arguments for the immortality of the soul, Lamont's book reviewed current thinking on the possible immortality of the body. Death, the author explained, was neither 'necessary' nor 'universal'. It had first appeared 'only after living things had advanced some distance on the path of evolution'. There were and are, as Metchnikoff had pointed out, various single-celled organisms which in effect never die: reproducing by division, Lamont explained, 'one individual becomes two, with nothing left behind in the process corresponding to a corpse. . . . Thus the whole chain of animal life, reaching through hundreds of millions of years, is really one unbroken continuum of deathless protoplasm.' Referring to Carrel's work, Lamont explained how single cells, isolated from the larger organism they belonged to, were 'preserved from death indefinitely . . . It seems that the individual cells in a body do not die because mortality is inherent in their structure, but because they are parts of a very complicated system which is constantly making very great demands on them and of which one section, small or large, sooner or later lets the others down.'[59]

What makes human life what it is – the mass congregation of millions of cells – ultimately brings about its death. As Lamont concluded:

> Biology does not strictly rule out immortality for human person-alities, but it insistently indicates that any immortality must be based on natural bodies. It is not inconceivable, however, that at some far-off day science should learn how to prolong indefinitely the life of human bodies.[60]

§

It was in Chicago, where Carrel had first worked on coming to America, that some of the most interesting research in tissue transfer was going on. In 1914, Dr George Frank Lydeston, Professor of Genito-Urinary Surgery and Venereal Diseases at Chicago's College of Physicians and Surgeons, carried out human testis transplants into patients with various degenerative conditions. In 1920, he told the *New York Times* that he had been able to retard senility, cure grey hair, and reverse defective sexual development. He even apparently trans-planted a human testicle onto his own body, informing a colleague that it had invigorated him greatly.[61] The Chicago researchers inspired others in the USA: in October 1919 Dr Leo L. Stanley, a young medical officer at San Quentin Prison, California, removed the testicles from the corpse of an executed prisoner and transplanted them into a prematurely aged sixty-year-old inmate.[62] Dr Stanley went on to carry out a number of similar transplants, though with human matter in limited supply he also experimented with animal testes. By 1922, he would publish an article in the medical journal *Endocrinology* titled 'An analysis of one thousand testicular substance implantations'. American doctors in other prisons attempted similar rejuvenation experiments. The reported effects were encouraging, and in 1920 the former middleweight champion boxer Frank Klaus (1887–1948) became the first of many celebrities to announce he was having a testis graft.[63]

The doctor whose name became intimately associated with the use of animal hormones to rejuvenate and prolong human life was Serge Voronoff (1866–1951). Son of a Jewish Russian vodka manufacturer, Voronoff emigrated to Paris when he was eighteen, and studied classics before taking up medicine. Graduating in 1893, he became a French citizen in 1895 and spent some years living in Egypt, where he was personal physician to the court of the Khedive Abbas II. Returning to Paris in around 1910, Voronoff started experimenting with tissue transplantation.[64] It was, as we have seen, an exciting field. The discovery of antiseptic techniques by the English surgeon Joseph Lister in the 1860s, together with advances in anaesthesia, had greatly extended the boundaries of surgery. By the early years of the twentieth century unsuccessful attempts were even being made to transplant animal organs into humans. Voronoff's aims were less ambitious, though they proved more lucrative. He started off working on the transplantation of ovaries from sheep in an apparent attempt to devise a cure for infertility. He corresponded with Alexis Carrel, and it is possible that the two men met when they were drafted into the French army's medical service during the Great War.[65] Voronoff would subsequently claim that he had spent nine or ten months working with Alexis Carrel at the Rockefeller Institute before the war – an assertion the institute denied.[66]

As a military surgeon dealing with maimed casualties of battle, Voronoff made successful bone grafts, including ones using monkey bone. Discharged from the army on medical grounds in 1917, he was appointed to the biological laboratory of the Ecole Pratique des Hautes Etudes, eventually becoming director of the physiological station at the

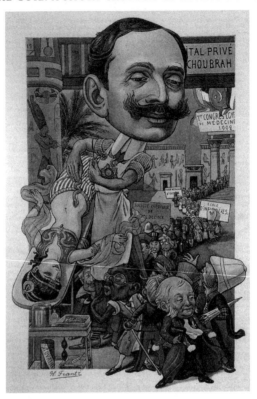

Figure 5.2 'Serge Voronoff performing an appendectomy': caricature by H. Frantz from *Chanteclair*, 1910, no. 59. The son of a Russian vodka manufacturer, Serge Voronoff enjoyed success as personal physician to the court of the Egyptian king, before he made his name as the promoter of life extension through the grafting of monkey glands. This unusual caricature depicts him undertaking an operation at the Congress of Tropical Medicine in Cairo in 1902.

Collège de France in Paris, where Brown-Séquard had once worked.[67] With the financial and practical assistance of a very wealthy American oil heiress, Evelyn Bostwick, Voronoff was soon experimentally grafting slices of testis gland taken from young rams and goats onto the testicles of older animals: he stated that these had a rejuvenating effect. He also grafted them onto castrated rams, claiming that their sexual activity returned.

There must have been major defects in Voronoff's experiments: without the aid of modern drugs, the grafts would have been rejected within days. Furthermore, Voronoff's observational methods were flawed: he lacked a control group and made no actual measurements (such as changes in weight or physical activity), only objective visual observations and took photographs.[68] As Dr Walter R. Hawden, editor of the British anti-vivisectionist journal *Abolitionist*, suggested (probably correctly) in October 1919, Voronoff had in all likelihood mistaken 'the excitement of fever, due to the decomposition and absorption of the constituents of the transplanted gland, for the liveliness of rejuvenation'. He further pointed out that the Rockefeller Institute had 'warned people not to be led away by Dr Voronoff's confident claims'.[69]

Nevertheless, in his 1920 book *Life: A Study of Methods to Revive Vital Energy and Prolong Life*, Voronoff pointed to Alexis Carrel's work, which suggested that living cells cultivated outside the body were effectively immortal. This suggested to Voronoff that senescence was due either to poisons or to some physical deficiency which occurred only in more complex organisms. He located the source of this supposed deficiency in the testes and the decline in the production of semen. Human eunuchs and gelded horses, he pointed out, lose their virility and age prematurely, whilst long-lived men, such as Goethe and Thomas Parr, possessed well-preserved testes.[70] It was weak and inaccurate evidence, but it satisfied Voronoff.

Almost inevitably, the popular press focused on the prurient side of Voronoff's work. He and Evelyn Bostwick (who in 1920 became his wife, and soon made him a very, very wealthy man)[71] were figures of fun in Europe and the USA. But, as Evelyn told the *New York Times* journalist and physician Dr Van Buren Dorne in August 1920, using monkeys to enhance flagging human sexual energies was not their objective: 'Our aim is to avert premature old age,' she explained, 'and to prolong life.'[72] She told Dorne that, despite reports to the contrary in the press, they had carried out their procedure on only one human subject; they had been forced to use glands from a live chimpanzee because French law forbade the 'mutilation' of human cadavers.[73]

One of the Voronoff's first human subjects was an enthusiastic seventy-four-year-old English volunteer named Arthur Evelyn Liardet. He underwent a free operation on 2 February 1921: both testicles of a chimpanzee were removed, each one being cut into six pieces. Three pieces were then grafted on to Liardet's testicles. (Each chimpanzee thus provided sufficient material for at least two operations, and sometimes Voronoff grafted matter onto only one testicle.)[74] Liardet was highly satisfied with the result. In October 1922, the *New York Times* reported that the septuagenarian Englishman – who produced a birth certificate to prove that he had been born in October 1846 – had 'the appearance and physique of a man of about 55'. Liardet explained to the paper that Voronoff had told him: 'when I again felt myself growing old he would repeat the operation, and that he could perform it, in all, three times. That ought to take me to the age of 150.'[75]

Photographs of Arthur Liardet 'before and after' appeared in Voronoff's 1925 book *Rejuvenation by Grafting*. Van Buren Dorne, the sceptical New York doctor who had interviewed Voronoff's wife in 1920, found them 'most striking'. Whilst the first showed 'the disappointed and dissolute countenance of a broken old man', the three pictures of Liardet taken after his operation 'show a man becoming progressively younger, obviously enjoying a self-satisfied and insolent triumph in the recrudescence of health and vigor'.[76] In 1925 the French politician Maurice Lebon predicted that monkey gland transplants would soon cease to be a 'laughing matter', and that, once accustomed to the idea, people would consider the operation 'a natural remedy for old age'. According to the *New York Times*, a study by the Collège de France into 300 operations indicated that 'the results have been all that was hoped for', and Dr Lebon advocated an immediate increase in the breeding of chimpanzees in French Guiana.[77] By the end of 1926 Dr Voronoff claimed he had performed around a thousand chimpanzee-to-human grafts.[78]

Was it Thomas Parr's century and a half that marked for Voronoff the true potential of human longevity? In 1927, he told the British and American press in Paris that his work with animals and humans suggested 'that human life could be extended to 125 years and senile old age practically eliminated'.[79] When he visited England the following year, he declared at a reception held in the home of a London surgeon: 'The human body is so made, that it is capable of lasting 150 years, but people grow weak and then pneumonia cuts them down. My treatment makes them far more able to withstand illness.'[80] In an interview with the *Daily Express*, he confidently predicted that in fifty years' time 'human beings should live twice the normal span'. He foresaw, he added, 'a world peopled by a race of supermen'.[81] At the height of his fame he would be dubbed 'the Great Rejuvenator'.[82]

Voronoff's work with animals was not universally well received. A meeting of the Animal Defence and Anti-Vivisection Society in London issued howls of protest during a visit he made to England in 1928. What they called the Voronoff 'cult' had not only brought about the gland operation between humans and chimpanzees, but glands were even being taken from corpses. The practice, they declared, was 'an offence against morality, hygiene and decency'.[83] The Liberal MP Frank Briant took up the attack in the House of Commons. He declared 'that the British public were extremely anxious that Dr Voronoff should not have an opportunity of conducting his dangerous and disgusting operations here'. The Home Secretary carefully explained that such experiments were not permitted in Britain.[84] Voronoff, anyway, was unmoved. He claimed that Britons were 'flocking' to Paris to have the gland operation, and that he had some 300 trained surgeons around the world as 'disciples'.[85] He eventually established himself on the French Riviera, where he kept a zoo of male and female monkeys and continued implanting their glands into aged men and women. In 1933 he would be made an Officer of the Legion of Honour, in recognition of his work.[86]

In 1926, however, Van Buren Dorne noted that Voronoff's rejuvenation procedures had been 'condemned' as 'useless and even harmful' in a recent editorial in the *Journal of the American Medical Association*.[87] Through the late 1920s there was growing scepticism: in 1928 a German doctor working at the university hospital in Freiburg reported that animal glands grafted onto humans were quickly absorbed or atrophied 'beyond their power of function'.[88] Two years later, a group of eminent Austrian physicians completed a prolonged trial of gland transplantation; they reported that it was 'effective only in a relatively small number of cases', and even then 'only for short periods'. In only one patient, a seventy-four-year-old man, was a 'lasting improvement' achieved, and even he 'failed to exhibit a regrowth of hair, disappearance of wrinkles or other striking signs of rejuvenation described by Dr Voronoff'.

But it must be noted that scientists did not reject the concept of gland transplantation outright. Their conclusion 'was that only aging, not aged, organisms were capable of regeneration'.[89] The procedure could not turn back the clock – though it might slow down the hands. Perhaps it worked for Voronoff: at seventy, he married his third wife, a young woman of only twenty.[90]

At the same time as the monkey gland phenomena, other surgical rejuvenation techniques were taking off. Although surgical operations had been undertaken on noses, chins and ears for aesthetic reasons in the nineteenth century, the first account of aesthetic surgery aimed specifically at ageing occurred in the first decade of the twentieth. In 1901, according to his much later claim, the surgeon Eugen Holländer (1867–1932) undertook the first rhytidectomy or 'face-lift'. Holländer wrote that one of his clients, a Polish aristocrat, suggested the procedure and persuaded him to carry it out.[91] The Great War, in which many men were horribly maimed, had helped advance techniques of reconstructive surgery – processes in which Voronoff first made his name. In 1926, the 'beauty surgeon' Charles Willi would publish *Facial Rejuvenation: How to Idealise the Features and the Skin of the Face by the Latest Scientific Methods*. Willi, who used subcutaneous fat injections, explained how a female patient would enter his studio 'with a permanent frown' and leave as 'a young unwrinkled woman. There is no pain,' he assured his readers, 'there is no danger, it is an instantaneous and painless rejuvenation.'[92]

These surgical techniques, however, have nothing to do with the actual *prolongation* of life.[93] The scientists I am discussing here really believed they could make people *live* longer, not just *look* younger.

§

Serge Voronoff's principal competitor in the inter-war quest to prolong life was the Viennese physiologist Eugen Steinach (1861–1944), professor of biology at the University of Vienna and director of the city's Biological Institute of the Academy of Science. Though the two men and their processes were (and still are) sometimes confused with one another, Steinach stated bluntly in 1923: 'Voronoff is not to be taken seriously.'[94] Raymond Pearl thought likewise.[95]

Steinach came from a medical family and had worked longer and more diligently in this field than Voronoff. As early as 1894, he was vivisecting animals, in particular frogs, rats and guinea pigs, and perceiving apparently close connections between the sex glands and the ageing process.[96] In 1902, two anatomists, Pol André Bouin (1870–1972) and Paul Ancel (1873–1961), experimented with ligating the vas deferens in rabbits – an operation now known as a vasectomy. They noted that many of the ligated animals appeared to show increased sexual activity. Steinach investigated the effects of vasoligation in bulls, rams and rats, with what seemed like encouraging rejuvenating results. During the war he studied a soldier whose testicles had been destroyed in battle, whilst in 1918 his friend, the urologist Robert Lichtenstern, undertook vasectomies on three old or prematurely aged men – with apparently positive effects.[97] As Steinach later wrote, 'reactivation' was not simply about 'supporting a few decaying branches'; rather, it attacked 'the ageing process at its root'. Sex was 'the most obvious root, because it is the root of life. Just as it produces physical and psychic maturity, induces and preserves the period of flowering, shorter or longer, here richer, there poorer, so it is also responsible for the withering of the body and gradual loss of vitality.'[98]

Figure 5.3 Eugen Steinach: photograph by J. Scherb, after a painting. The Viennese physiologist Eugen Steinach conducted vasectomies on rats, rams, bulls and men over a number of decades, and reported apparently startling rejuvenatory effects. In 1923, Dr Peter Schmidt predicted that 'indefinite prolongation of life by a series of Steinach operations is well within the bounds of possibility'.

In 1920, Steinach published *Rejuvenation through the Experimental Revitalization of the Aging Puberty Gland*: 'almost overnight', according to a 1945 report, he 'became one of the most talked-of men of science all over the world . . . praised as a genius by some and condemned as a quack by others'.[99] The book was followed in 1922 by a documentary film made by the Austrian government, explaining his research and showing how the Steinach operation resulted in rejuvenation and improved sexual prowess. As Raymond Pearl put it in his interested, if sceptical, 1922 assessment, Steinach claimed 'to bring about in highly senile animals a great enlargement of all the sex organs, a return of sexual activity, previously lost through old age, and . . . a resumption of the conditions of full adult vigor . . . together with a considerable increase in the total duration of life'. Pearl, however, found that the 'presumption that Steinach's experiments have really brought about a statistically significant lengthening of life is large, and the basis of ascertained fact small'.[100] But Steinach's colleague, Dr Peter Schmidt, predicted in a 1923 address given in Berlin and reported in *Time* magazine that 'indefinite prolongation of life by a series of Steinach operations is well within the bounds of possibility'.[101] By the early 1920s Steinach had gained international recognition.

Clearly the devastation of the Great War played a part in the enthusiastic reception. As Dr Schmidt explained, 'When the last gun was silenced, not only were the best men of every nation taking part in it decimated, but the survivors seemed to be absolutely at the end of their resources of endurance.'[102] Steinach seemed to be offering new hope to a war-weary world. In the 1920s – an era that celebrated youth and the Weimar cult of outdoor vitality, strength and beauty – his theory had obvious popular appeal.

Though Raymond Pearl believed Steinach had 'not proven that any really significant lengthening of the life span has occurred' in his operations on rats,[103] the Austrian physiologist's colleague, the German-born physician Dr Harry Benjamin, carried out numerous Steinach operations in New York. Benjamin later reported that Vienna in the 1920s saw 'well over a hundred university professors and teachers', including Sigmund Freud, have 'the platform of their efficiency' extended by this rejuvenation or 'reactivation' method. According to Benjamin, Freud – who had known Steinach when they were students – hoped it might slow the recurrence of his jaw cancer.

He also hoped it might improve his 'sexuality, his general condition and his capacity for work'. Benjamin stated in 1945 that Freud was 'emphatic' that the operation had worked – and Freud also expressed interest in Steinach's attempts to 'cure' homosexuality by replacing a homosexual man's testicle with one taken from a heterosexual.[104]

'Rejuvenation cannot make immortals', Steinach's ill-fated Viennese colleague, the zoologist Dr Paul Kammerer, counselled in 1924. But now that 'the principles of rejuvenation' had been made clear, he felt 'it would scarcely seem magical if the perfections of the first discovery along these lines would become more numerous from day to day – growing immeasurably'.[105] In 1926, an American journalist estimated that over eight thousand men had been 'Steinached' in the United States, and 'perhaps three times that number in Europe'.[106] As another American journalist asked: 'Does Professor Eugen Steinach bring us

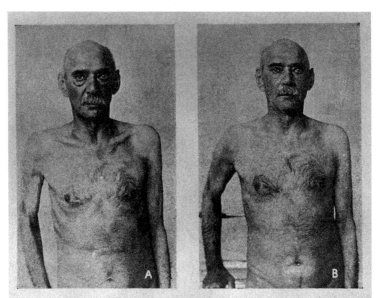

(a) SEVENTY YEAR OLD PATIENT BEFORE THE STEINACH OPERATION
(b) TWO MONTHS AFTER THE STEINACH OPERATION
(*Ufa Steinach-Film*)

Figure 5.4 Before and after a Steinach operation: photograph from *Rejuvenation and the Prolongation of Human Efficiency*, by Dr Paul Kammerer (1924). This 'before and after' image illustrates the apparent positive effects of a Steinach operation. Kammerer committed suicide in 1926, following accusations that he had falsified evidence in his evolutionary studies of midwife toads.

the knowledge that the serpent promised Eve, that shall make us like gods?' Looking to future discoveries, he pondered: 'If we halt the insidious advance of Age, may we not, in time, challenge Death?'[107]

Encouraged by this reception, Steinach developed a version of his operation for women. Alarmingly, it involved irradiating the ovaries with X-rays. The American author Gertrude Atherton visited Benjamin's clinic in 1923, and quickly recorded her experience in her bestselling novel *Black Oxen* (which became a silent movie, featuring Clara Bow). Atherton considered the X-rays' apparently positive results as 'proof that the effects of old age can be deferred'. She later claimed that after a month of treatment her creative faculties were restored, and that she had done some of her 'best work' in the decade since.[108] Atherton claimed that, as a result of her novel, Benjamin was 'besieged' by women seeking the operation; whether many actually underwent it is not so clear. But Atherton undertook further radiating operations, and to the end of her long life took hormone pills prescribed by Benjamin. She died in San Francisco in 1948, at the age of ninety, having published over sixty books.[109]

Dr Benjamin also lived long, dying in 1986 at the age of 101. By then he acknowledged that the apparently positive effects of these operations were based solely on 'autosuggestion' – the placebo effect. As early as 1927, Morris Fishbein, editor of the *Journal of the American Medical Association*, had dismissed Steinach's claims for rejuvenation, complaining: 'The records of the surgeons who do the Steinach operations are without scientific controls.'[110] But, like Voronoff, Steinach should not be seen as a charlatan: between 1921 and 1938 he would be nominated by colleagues six times for the Nobel Prize – though he actually never won it.[111]

Furthermore Steinach's work attracted the attention of reputable physicians. They included the Australian doctor Norman Haire, who ran a successful practice in London's prestigious Harley Street. In 1923 Haire explained to the sexologist Havelock Ellis how at first he had been 'very sceptical of the claims made by Steinach and his co-workers'. But he had since seen 'some amazing results', and told Ellis he could 'no longer doubt that, in some cases at least, Steinach's operation brings about a real rejuvenation'.[112] Haire performed his first Steinach operation in 1922 and, having visited Steinach's impressive laboratory in Vienna, in 1924 he published *Rejuvenation: The Work of Steinach,*

Voronoff and Others. Haire pointed out that this treatment was no panacea: it could not cure cancer, for example, and in some severe cases there would be no impact at all. But, if the subjects were carefully chosen, Haire believed that the operation could be 'very useful'. It could lower high blood pressure, increase muscle energy, 'cause improved nutrition of skin and renewed growth of hair; improve power of concentration, memory, temper, capacity for mental work; and possibly increase of sexual desire, potency, and pleasure'.[113]

Haire explained, however, that it was hard to give a simple 'yes' or 'no' answer to the question of whether Steinach's operation actually prolonged life. This was 'because it is impossible to know what age any individual would have lived to without the operation'. But one of Steinach's ligated rats had lived for thirty-seven months, whilst its siblings – which had not received the operation – had died at the usual age of about twenty-eight months. This 30 per cent increase in longevity, Haire felt, 'seems pretty conclusive'. But he warned that it was still 'too early to decide' if the same effects might happen with humans.[114] He acknowledged that the sudden death of one of Steinach's septuagenarian patients shortly before he was to deliver a lecture entitled 'How I was made twenty years younger' was unfortunate. But, as Haire explained, this man had been 'too confident in his renewed vigour'. He had foolishly 'overtaxed his powers', and had died 'from an attack of angina pectoris, a disease from which he had suffered for many years before the operation'.[115]

One of the most prominent converts to Steinach's operation was the Irish poet William Butler Yeats (1865–1939). Early in 1934 he read the English translation of a recently published book by another of Steinach's supporters, the surgeon Peter Schmidt.[116] In *The Conquest of Old Age: Methods to Effect Rejuvenation and to Increase Functional Activity*, Schmidt confidently declared: 'We are now able to regard old age as an illness, as a result of decay in the reproductive gland.'[117] Yeats was then sixty-eight and had been ailing for some years. He was probably impotent.[118] He visited Norman Haire in London, who undertook a Steinach operation, followed by a series of injections, possibly of a sex hormone.

To determine whether or not the operation had succeeded, Haire conducted an unusual experiment. He invited Yeats and a young writer, Ethel Mannin, to join him for dinner. Mannin, who was then thirty-four, later recalled how Haire asked her 'to put on my most alluring evening dress and all my sex-appeal, for the famous poet to test out his rejuvenated reaction to an attractive young woman'. Although the dinner was a failure, Yeats and Mannin did go on to have a love affair. Whether this was consummated is unknown; but in the 1960s Mannin destroyed the letters relating to what she called 'Yeats's rejuvenation operation', telling her daughter that she found them 'quite disgusting and revolting'.[119]

Yeats's medical friend, Dr Oliver Gogarty – who complained that Yeats had not consulted him before submitting 'to that humbug Steinach' – wrote that after the operation the aged poet was 'trapped and enmeshed in sex'; Gogarty even worried that he had gone 'sex mad'.[120] Dublin newspapers (confusing Steinach's operation with Voronoff's) mockingly dubbed Yeats 'the gland old man'.

But Yeats was satisfied with what he would call his 'second puberty'. As he told his wife in January 1935, 'Norman Haire has done me much good, I expect increased working powers etc.'[121] She in turn would tell the biographer Richard Ellman that the effect of the Steinach operation on the poet's mind in his last years had been 'incalculable'.[122] As well as having at least four sexual (if not fully consummated) relationships between the operation and his death in 1939, Yeats produced some of his finest poetry. A month before his death, he wrote 'I am happy, and I think full of an energy, of an energy I had despaired of'.[123] Haire might not have much prolonged the poet's *physical* life; but he undoubtedly extended his working life. And all from a medical procedure that had absolutely no capacity to achieve its intended aim! Certainly, as Godwin had believed, the power of the mind over the body is not to be underestimated.

Yeats was a relatively late convert to Steinach's operation. Though the Swiss physician Paul Niehans (1882–1971) would successfully promote expensive cellular rejuvenation therapies well beyond mid-twentieth century,[124] in 1931 Alexis Carrel declared in an article in *Science* that 'no senescent organism has ever been rejuvenated by the procedures of Steinach and Voronoff. So far, the process of aging remains irreversible.'[125] The British physiologist Samson Wright concurred: testicular grafts, he observed in 1936, might stimulate physical activity, but 'they cannot restore the worn heart, arteries and essential organs to their normal state'.[126]

In 1936, Steinach complained that, from the start, the 'generalized effects on the body and the mind' of his operation had been ignored: 'Reactivation was besmirched by the odium of sexuality.'[127] Steinach died in Switzerland in 1944, exiled from his native Austria by Hitler's invasion. He was eighty-three and, according to Harry Benjamin, a 'lonely, uprooted, and somewhat embittered man'.[128] His rival, Voronoff, also died in Switzerland – a year older, and seventeen years later. As a Russian-born Jew, he had been fortunate to escape to America when France fell to the Germans in 1941. Three, perhaps four of his brothers were not so lucky, and were probably murdered by the Nazis. Voronoff would be described in a 1952 review of his career as a 'broken and almost forgotten man'.[129]

§

Despite the fashion for glandular therapies in the inter-war period, evolutionary ideas on the prolongation of life did not disappear. The same question that had troubled Darwin and Galton still puzzled scientists: were humans evolving? As Edwin Conklin, Research Professor of Biology at Princeton University, observed in his 1921 book *The Direction of Human Evolution*:

> The prolongation of individual human lives by means of medicine, surgery, and general scientific knowledge has led many persons to hope that the present maximum length of life may be greatly extended in the future so that men may once more reach the reputed ages of the patriarchs. But the saving of individual lives has not extended the maximum length of life. The oldest individuals to-day are no older than those of prescientific times. The *average* life of the race has been lengthened chiefly through the reduction of infant mortality. But since it has been proven that longevity is hereditary, it may well be that the artificial prolongation of the lives of the hereditarily weak and short-lived may actually reduce the natural longevity of the race as a whole.
>
> In any event there is no probability that science will greatly extend the present maximum length of life, and there is no basis whatever for the hope which is sometimes expressed that it will ultimately banish death altogether.

Without death, Conklin pointed out, 'and the succession of generations, there could be little or no evolution'. Thus he felt that 'under present conditions immortality of the body would be the greatest possible hindrance to human progress'. He concluded with the observation that there were 'no indications that future man will be much more perfect in body than the most perfect individuals of the present', nor than the most perfect men and women of ancient Greece.[130]

Other scientists, however, disagreed. In 1908 the Scottish-born inventor and eugenicist Alexander Graham Bell (1847–1922) had begun a statistical study of longevity and human heredity.[131] Bell, who had been a Darwinist since his early twenties, had emigrated to North America in 1871. He felt that the eugenics movement in the USA focused too strongly on 'negative' rather than 'positive' methods of improving the human stock. As he pointed out in his article 'On the possibility of improving the human race', published shortly before the start of the Great War, '[p]reventing the marriages of dwarfs would not produce giants. There would be no advance in stature as a result of such a process.'[132] Such negative preventative measures simply would not work in perfecting the human species. Improving longevity required *positive* action, such as the orchestrated intermarriage of couples from known long-lived families.

Bell's source supporting his argument that longevity was hereditary was Reuben Hyde Walworth's *The Genealogy of the Hyde Family* (1864). Described by Raymond Pearl in 1920 as 'one of the finest examples in existence of careful and painstaking genealogical research',[133] this book included the names of some 8,797 individuals who were either members of, or linked to, the Hyde family. It also included as many of their vital dates (including births, marriages and deaths) as Walworth had been able to discover. Aided by two assistants, Bell used this study to determine whether there was any correlation between the length of an individual's life and that of their parents. He also allowed for other variables that might have influenced longevity, such as the number of offspring and siblings, the parents' ages when an individual was born, etc. His statistical study took six years to complete; the results were published in 1918 as *The Duration of Life and Conditions Associated with Longevity.*

According to Bell's analysis of Walworth's records, when both parents lived to eighty or beyond, their children's mean length of life of was 52.7 years, with approximately 21 per cent themselves reaching the age of eighty. But when both parents died before sixty, their offspring's mean length of life was only 32.8 years, with only 5 per cent living to eighty. Bell was well aware that a variety of other factors (such as an early death in war or by accident) played a part in determining the length of any individual life; but from his researches he concluded that heredity was 'deeply involved in the production of longevity'.[134]

In an article for the *National Geographic Magazine*, titled 'Who shall inherit long life?', Bell explained that it was 'not longevity itself that is transmitted, but something else that tends to produce long life. What is really inherited is probably a tough, wiry constitution, which enables the fortunate possessor to survive the multitudinous ills that flesh is heir to and live on to the extreme limit of human life.'[135] Though he was not a mathematician by training, and there were flaws in his methodology, subsequent studies have supported Bell's main conclusion: there *is* an hereditary element to longevity.[136]

Bell was not interested only in compiling statistics, however. As an active eugenicist, he wanted to see this knowledge put into action. So in 1914 he founded the Genealogical Record Office for the Collection and Preservation of Genealogical Records Pertaining to Long Life. Based in Washington DC, the GRO's aim was to provide a source for the names and biographical details of people who had lived to eighty or beyond, and to authenticate them. But there was another motive. As Bell wrote in a notebook in April 1918, 'I believe that the publication of a mere list of very aged people will indirectly stimulate marriage among their descendants.'[137]

He was not alone in his ideas on hereditary longevity. As far as Julian Huxley (1887–1975), Professor of Zoology at King's College, London, was concerned, since longevity was 'clearly in part inheritable' in animals, he had 'no doubt that this holds good also for man'. By 'eugenic measures', he predicted in a 1926 essay, 'we could unquestion-ably raise the average span of human life, even without further progress in hygiene'.[138] The *New York Times* reported the suppositions of this essay rather excitedly under the headline, 'Huxley sees life prolonged in future: Eugenic measures, as well as hygiene, relied upon by biologist to extend man's span'.[139] In 1932, Julian's younger brother, the writer Aldous Huxley (1894–1963), published his novel *Brave New World*. By employing a variety of measures, including gonadal hormones and the transfusion of young blood, his future society 'abolished' all 'the physiological stigmata of old age . . . at sixty our powers and tastes are what they were at seventeen'.[140] Huxley followed *Brave New World* with *After Many a Summer* (1939), in which an ageing Californian millionaire funds researchers to seek new methods of rejuvenation. It is eventually discovered that the 200-year-old Fifth Earl of Gonister and his mistress have found a virtual immortality by eating raw the guts of a rare species of carp. Unfortunately, however, the aged earl has degenerated into an ape.[141]

Not only novelists enjoyed such flights of fancy. In 1930, in the conclusion to an essay on 'The purposive improvement of the human race' in which he discussed eugenics, euthenics (the way in which improvements in living standards and environment lead to benefits in health) and the possibilities of future progress, the abovementioned Edwin Grant Conklin of Princeton University observed optimistically:

> By his knowledge and power man has in a measure risen above nature, he has eaten of the fruit of the tree of knowledge and has become as the gods, knowing good and evil, and now it remains to be seen whether in future ages his race may secure the fruit of the tree of life and become immortal.[142]

With their grand possibilities of human improvement, of breeding a race of 'supermen', eugenics and euthenics were of keen interest to numerous respectable men of science whom we have already encountered. As well as carrying out celebrity vasectomies, Norman Haire was a member of the Eugenics Society; Raymond Pearl was a keen breeder of animals and had no doubt that, given 'omnipotent control', it would be possible to breed within a few generations a race of men 'considerably superior' to the average.[143] And his colleague Alexis Carrel became an infamous supporter of eugenics.

In 1935 Carrel published a curious book, *Man, the Unknown*. It quickly became a bestseller: within a year it had been translated into thirteen languages, including his native French.[144] The book included not only a summary of recent research into medicine and physiology, but also Carrel's own reflections on the decline of modern civilization, mixed together with his rather mystical religious ideas. The final chapter, 'The remaking of man', explored his eugenic beliefs. These were based largely on 'positive' methods that encouraged the improvement and reproduction of the fit. But he also advocated some negative measures, including ones that would prevent 'defectives and criminals' from reproducing. In a notorious passage, Carrel suggested society should find economic ways to 'dispose' of undesirables:

> Those who have murdered, robbed while armed with automatic pistol or machine gun, kidnapped children, despoiled the poor of their savings, misled the public in important matters, should be humanely and economically disposed of in small euthanasic institutions supplied with proper gases. A similar treatment could be advantageously applied to the insane, guilty of criminal acts.[145]

Despite this intimation of lethal gas chambers, Carrel mostly advocated 'positive' action to improve the stock of mankind. A variety of measures proposed in his posthumously published *Reflections on Life* included attention to diet, plenty of exercise, fresh air, and avoidance of the trivial materialism of modern, industrialized existence (such as not driving around in fast cars, listening to the radio, reading newspapers or visiting cinemas). Together, these acts would make us fitter, stronger. Too many people, he complained, had allowed themselves 'to grow feeble'. Eugenics, therefore, was an 'indispensable virtue if we are to save Western civilization'.[146]

As Carrel pointed out in *Reflections on Life*, the 'progress of hygiene' had 'not made us live longer'; it had 'merely increased the average duration of life'. Actually to prolong our lives and 'increase the intelligence of the race', we would 'have to find the secret of speeding up the natural march of evolution'. He suggested that the moment might have come 'for scientists to see whether it is possible to modify the quality of the brain matter and of the endocrine glands in such a way as to improve the mind. Perhaps one day we shall be able to make great men just as bees make queens.'[147] Carrel warned, however, that, even if through new discoveries in science 'we came to the point . . . of periodically rejuvenating ourselves and even of prolonging our lives for two or three centuries, death would not be overcome because the structure of our body makes death a necessity'.[148] His own scientific discovery, the 'immortal cell', was not sufficient evidence for Carrel to believe that death, the ultimate law of nature, might one day be conquered.

But how was such a 'speeding up' of evolution to be achieved – other than by positive eugenic means? In *Man, the Unknown*, Carrel advocated the establishing of a new international institution of medical researchers. Based along the lines of the Rockefeller Center in New York, it would work wholeheartedly towards the improvement of mankind, over many, many years. Such a project would require, he explained, 'an intellectual focus, an immortal brain, capable of conceiving and planning its future, and of promoting and pushing forward fundamental researches, in spite of the death of the individual researchers'.[149] It sounds a lot like what Bacon had proposed in his *New Atlantis*.

A crushing review of *Man, the Unknown* appeared in the *American Sociological Review* in 1936. According to its author, Read Bain of Miami University, Ohio, Carrel's solution to societal problems was a mix of 'pure poppycock, naïve speculation, opinionated guess, [and] biological determinism mingled with medieval-minded mysticism'. It was 'a kind of research-foundation Fascism and superior-manism which sneers at democracy'.[150] Bain was right. With hindsight we can pinpoint exactly where well intentioned ideas such as Carrel's were leading, as accurately as a train track leading to a Polish death camp.

It was, of course, in fascist Germany that eugenic ideology was carried to its logical and appalling conclusion. Nazi techniques of mass-murder were first practised on those whom the regime defined as insane, delinquent, asocial or degenerate, before being perfected on an industrial scale against continental Europe's Jewish population. In 1935 (the same year Carrel's bestseller was published, and two years after the newly elected Nazi government had introduced its first sterilization legislation), a law for 'the protection of the hereditary health of the German people' was passed, requiring all those planning to marry to obtain a 'certificate of fitness'. These licences could be refused on grounds of 'hereditary illness' or contagious disease, and were aimed at excluding the 'racially less valuable' from the 'national community'.[151]

Also in 1935, Heinrich Himmler established the *Lebensborn* ('fountain of life'). As a new unit of the SS's Race and Settlement Office, it had the explicit 'positive eugenic' aim of improving the stock of Aryan blood. This was to be achieved by raising and 'integrating' stolen children of conquered peoples who, as individuals, appeared to be genetically superior. The more numerous genetically 'inferior', mean-while, would be destroyed: 'you may call this cruel,' Himmler told an audience in 1943, 'but nature is cruel.'[152] Meanwhile, in experiments echoing Steinach's research with guinea pigs, SS doctors experimented with 'curing' homosexual concentration-camp inmates by injecting them with male hormones.[153]

The Nazi eugenic programme, of course, was more interested in youth, vitality and 'racial purity' than in the prolongation of life. Indeed, there is something abject about using that phrase in the context of an obscene political ideology responsible for the premature deaths of millions of innocent people. In his essay 'Vererbung und Auslese' ('Heredity and selection'), Wilhelm Schallmeyer, a standard bearer of that 'race hygiene' ideology in Germany until his death in 1919, had disparaged the old as a burden on society; they would be 'blessed with an early death'. Though no racist, Schallmeyer's outlook helped set the tone for German approaches to eugenics in the first half of the twentieth century.[154] With the rise of Hitler, increasingly 'negative eugenic' laws were passed and programmes enacted in Germany and Austria, to the outrage of many eugenicists in Britain and the USA.

In March 1941, Alexis Carrel voluntarily returned from New York to Nazi-occupied France. With the assistance of the collaborative Vichy government, he established the French Foundation for the Study of Human Problems. One of its first projects was 'improving' the children of France: 'The quantity of children is insufficient', a foundation report explained in 1943. 'The same is true in general of their quality.' The foundation aimed to find ways 'to stimulate the birth of hereditarily well-endowed children'.[155] At the same time, in unconnected but not unrelated operations, the Vichy government and its representatives were helping the Germans to round up French Jews for deportation and (ultimately) termination.

Suspected of being a collaborator, Carrel would be spied on by the French Resistance; the Nobel laureate died in disgrace, one year after his country's liberation. When *Man, the Unknown* was reprinted a few years later, a reviewer declared that the 'chief value' of the book, which was 'seeded with the virus of racial superiority', lay in the fact 'that it will reveal to the reader why Carrel was a Nazi sympathizer'.[156] This is harsh on Carrel. He was no supporter of war, which, he wrote, 'resolves none of the fundamental human problems'. But he was a bitter and disappointed man. Human life, he wrote in his last years in France, 'is not a success . . . The means of destruction are progressing faster than the means of helping life. We have arrived at this singular moment of human history where we must either succeed or sink to our ruin in chaos and degeneration.'[157]

Sadly, Carrel had taken the wrong route at this supposed historical crossroads: genocide, not the improvement or prolongation of life, had been the result.

§

With the end of the Second World War and the discovery of the full horrors of the Nazis' 'Final Solution', we have advanced three centuries since Bacon posed the problem for modern medicine to overcome ageing. Science – as Bacon, Descartes, Condorcet and Franklin thought it would – was inching slowly towards a better understanding of senescence. But when was it going finally to solve the problem – and how many people would die waiting for the future to arrive? There was one possible answer: cold. Cold preserves flesh. Get it cold enough, and it might be possible to preserve a human body forever – or at least long enough for it to survive until medical advances catch up with that fundamental problem of life – death.

It took a while to get that far. In 1665, almost exactly forty years after Francis Bacon had died after stuffing a chicken with snow, Robert Boyle had observed that, 'notwithstanding Cold's being so important a subject, it has hitherto been almost totally neglected'. He dedicated a whole book to the topic. In *New Experiments and Observations Touching Cold*, Boyle discussed its effect as a preservative of 'inanimate Bodies'. He included an account of Swiss soldiers buried by an avalanche in the Alps, whose corpses, when the snow melted, were discovered 'in perfect condition'.[158] He likewise recorded that a man who had spent a winter in Scandinavia told him how 'in those colder Climates [they] preserve in Winter Dead Bodies unburied, and yet uncorrupted . . . three or four moneths together, or longer'.[159]

The following century, the Scottish surgeon John Hunter (1728–93) theorized that, if 'by freezing a person in the frigid zone' the blood could be stopped, 'all action and waste would cease until the body was thawed'. He speculated that, 'if a man would give up the last ten years of his life to this kind of alternate oblivion and action, it might be prolonged to one thousand years: and by getting himself thawed every hundred years, he might learn what had happened during his frozen condition'. He froze carp and then attempted to revive them. As he told his students: 'Like other schemers, I thought I should make my fortune by it; but this experiment undeceived me.'[160]

Gradually, science caught up with Boyle and Hunter. In 1899 the Scottish bacteriologist Allan Macfadyen of the College of State Medicine in London announced that, with the aid of the renowned chemist James Dewar (1842–1923), he had conducted experiments on bacteria frozen for some hours in 'liquid air' – a temperature approaching absolute zero. When carefully thawed and examined, Macfadyen found that '[i]n no instance . . . could any impairment of the vitality of the micro-organisms be detected'.[161] Two decades later, Professor D. Fraser Harris of Dalhousie University, in Halifax, Nova Scotia, conducted experiments by freezing frogs. He had heard that the Swiss scientist Raoul Pierre Pictet – the man who in 1879 had been one of the first people to freeze liquid oxygen – had claimed to have revived frogs frozen at –28°C. Harris found, however, that a frog could not endure temperatures below –10°C.[162] Though Harris explained in 1922 that what he called 'latent life' or 'suspended animation' could occur in humans in such states as narcolepsy, he averred that 'no human being can be dried up or frozen stiff like some of the lowlier creatures and yet live'.[163]

Yet the realization that cold could preserve bodies fundamentally intact inspired numerous fictional stories. As well as Mary Shelley's 'Roger Dodsworth: The reanimated Englishman', they included Louis Boussenard's 1890 novella, *10,000 Years in a Block of Ice*, and the American writer Neil R. Jones's 1931 short story, *The Jameson Satellite*. In Jones's story, a man's frozen body is preserved in a small spacecraft which orbits the earth for millions of years, before his brain is brought back to life by aliens.

Jones's science fiction inspired a young American, Robert Ettinger, to think about the possibility of freezing humans, and then resurrecting them, once medicine had advanced sufficiently to treat the causes of illness and death. Then, in the late 1940s, the French experimental biologist Jean Rostand successfully used glycerol to freeze frogspawn. Rostand's reports suddenly made Ettinger's idea seem plausible, and he waited expectantly for 'some prominent scientist to announce the arrival of the freezer era'.[164] It didn't happen. Death continued to be the end of life; the dead were buried or cremated, when to Ettinger it made clear sense that they should be frozen. By 1960 he was a teacher of physics at Wayne State University when, impatient, he sent letters 'to intelligent people', trying to interest them in this idea. He received some encouraging responses, but realized that a book was needed 'to supply a convincing explanation' of his theory.[165] So in 1962 he privately published *The Prospect of Immortality*, in which he outlined the science as well the moral, ethical, philosophical, religious and economic ramifications associated with freezing humans. This was followed in 1964 by a longer, commercial edition. Deliberately optimistic in tone, his intention, he declared in his book, was to show that 'immortality (in the sense of indefinitely extended life) is technically attainable . . . that it is practically feasible and . . . that it is desirable from the standpoints both of the individual and of society'.[166]

He opened the first chapter of this manifesto for life with a basic statement of how most people now alive 'have a chance for personal, physical immortality':

> *The fact*: At very low temperatures it is possible, *right now*, to preserve dead people with essentially no deterioration, indefinitely. . . . *The assumption*: If civilization endures, medical science should *eventually* be able to repair almost any damage to the human body, including freezing damage and senile debility or other cause of death.[167]

Certainly the science which convinced Ettinger that it would be possible to freeze and preserve humans had advanced considerably.[168] In 1956, Dr Harold T. Meryman of the Naval Medical Research Institute at the National Naval Medical Center in Bethesda, Maryland, had written in *Science* that '[u]nder any circumstances, storage in liquid nitrogen, at -197°C, can be considered as essentially indefinite'.[169] And Ettinger also quoted David MacDonald of Ottawa University, an expert in low temperature physics. In his 1961 book *Near Zero*, MacDonald had written that 'perhaps the day will come when, if you want it, you can arrange to "hibernate" for a thousand years or so in liquid air'.[170]

If proof was needed of what could be done, in September 1963 the *New York Times* reported that two babies had been born to women 'who had been artificially inseminated with sperm stored for two months at liquid nitrogen temperatures'.[171] The cryobiologist Audrey Ursula Smith of the National Institute for Medical Research in Mill Hill, London, was also reporting exciting results in 'supercooling' and 'reanimating' mammals, including hamsters, mice and rats.[172] In the 1940s she and Christopher Polge had discovered that, by using a medium containing glycerol, they could freeze spermatozoa and red blood cells without cell damage on thawing. At a 1958 conference sponsored by the New York Academy of Sciences, Smith even reported an experiment in which she had refrigerated monkeys until their blood froze; when revived, they had survived for as long as eighteen hours.[173]

As Ettinger pointed out, the research results reported in Smith's *Biological Effects of Freezing and Supercooling* (1961) were very important for his theory, since they suggested 'that mental faculties can survive freezing and thawing'.[174] Ettinger knew that the greatest impediment to his idea was not freezing the bodies. That part was now easy. The problem was thawing them afterwards.[175] But Jean Rostand, like Ettinger, was confident that 'the science of tomorrow' would find ways of repairing the cell damage which this invariably caused: 'When that happens, we shall have to replace cemeteries by dormitories,' Rostand predicted in his prefatory note to Ettinger's book, 'so that each of us has the chance for immortality that the present state of knowledge seems to promise.'[176]

Other recent research appeared to support Ettinger's second supposition: that in the future old age and death could be averted. At the National Medical Association in 1963, Dr Charles A. Brusch of Cambridge, Massachusetts, and Dr Murray Israel of the Vascular Research Foundation in New York, had reported favourable results from treating old people with thyroxine, a hormone of the thyroid gland. And, in what Ettinger called a 'sensational report', Dr Robert Wilson, a New York gynaecologist, had treated hundreds of patients with two female sex hormones, oestrogen and progesterone (along with diets, vitamins, minerals and exercise), to the great benefit of older women. With its echoes of Brown-Séquard, Voronoff and Steinach, Wilson would go on to promote the benefits of oestrogen in a bestseller optimistically titled *Feminine Forever.*[177]

Contemporaneous research not cited by Ettinger gave further support to his argument that the medical science of ageing had promising directions in which to advance. In *The Stress of Life*, the Austrian-born physician and endocrinologist Hans Selye wrote of his confidence 'that the natural human life-span is far in excess of the actual one'. What made him so certain was that, having performed 'quite a few' autopsies, he had 'never seen a man who died of old age. In fact,' Selye wrote, '*I do not think anyone has ever died of old age yet . . . We invariably die because one vital part has worn out too early in proportion to the rest of the body.*'[178]

Though Selye gave no indication of what he considered to be the maximum human span (he merely stated it 'could probably be greatly prolonged'),[179] even in the twentieth century stories continued to abound of extraordinarily long-lived individuals. In 1966, the magazine *Life* carried a photo story about a man named Shirali Muslimov, whom officials in the Soviet Union claimed was 160 years old; in the USA, an ex-slave named Charlie Smith died in 1979, supposedly aged 137 years. He was even included in the *Guinness Book of Records.*[180] Stronger evidence supporting Selye's opinion that we all die far too young appeared in a 1973 article in the *National Geographic* by Dr Alexander Leaf, Professor of Clinical Medicine at Harvard and Chief of Medical Services at Massachusetts General Hospital. The story Leaf investigated seemed plausible. He had travelled to Abkhasia on the eastern side of the Black Sea, an area renowned for the longevity of its inhabitants. These included Shirali Muslimov, the recent hero of *Life* magazine. By the time Leaf arrived in Abkhasia, Muslimov was dead. But the American doctor did meet a woman named Khfaf Lasuria, who claimed she was 141 years old. After making extensive investigations, Leaf concluded that Lasuria *was* close to 130.

Though the Soviet gerontologist Zhores A. Medvedev disputed all such claims of great longevity in an article in *The Gerontologist*, Leaf's *National Geographic* article, with its accompanying photographs, reached a wider audience.[181] As the myth of Thomas Parr had in earlier centuries, the present one also fuelled the impression that most of us age and die unnaturally young.

In such a cultural and scientific climate, it is not surprising that Ettinger's book made a big impact. To one academic reviewer, Ettinger was 'an utterly confused optimist', who had quoted his sources out of context and had grossly underrated 'the complications of biological structure'.[182] But Ettinger's proposition appealed to the burgeoning

New Age movement of the 1960s, and the idea inspired Woody Allen's 1973 movie *Sleeper*, in which a New York health-food fanatic is accidentally frozen and propelled two hundred years into the future.

As Allen famously quipped, 'I don't want to achieve immortality through my work. I want to achieve it by not dying.' Ettinger seemed to offer the best chance. This was not really death, he explained. The body should be thought of as 'not very dead': if frozen soon enough, Ettinger reckoned that 'the condition of most of the cells would not differ too greatly from that in life'. True, the body was not revivable by present medical methods, but that possibility would exist in the future.[183] This was an exciting new liminal state, a halfway house between here and oblivion. As a bet on immortality, Ettinger reckoned it came pretty cheap. He calculated that total start-up and running costs would be about $8,500 per body. This compared with around $1,000 for the average US funeral in 1964.[184] Not bad for a shot at immortality!

One of the first people cryonically frozen was Dr James Bedford, a seventy-three-year-old retired psychology professor from California. He died of cancer in January 1967 and was frozen in liquid oxygen in a 'dewar' by the Cryonics Society of California. The society explained to the press that, when a cure for cancer was found, Dr Bedford's body would be thawed, and an attempt made to revive him.[185]

Though Bedford's body is still frozen today, the take-off Ettinger predicted did not happen. Relatives were sometimes upset that their loved ones had not received a more 'normal' burial. Or they were suspicious. With start-up charges escalating to around $50,000 by 1975, and $5,000 a year for maintenance, this was big money to request from grieving relatives. The son of one interested subject complained that his mother had 'no way of knowing' whether it was just a scam to make money or a 'legitimate business'.[186] There continued to be a problem with trust. But, as one willing, dying 'volunteer' observed in 1990, the choice was between just one of two options: either 'to trust these people who look like they're coming from nowhere and have all these weird ideas' – or certain death.[187]

Although the technology has improved markedly since Ettinger first published his book – and although Ettinger is still alive, and still promoting the work of his company, the Cryonics Institute – freezing is not popular. It remains expensive: the US company Alcor charges $150,000 for a full body suspension, and $80,000 for a 'neural' – in which only the dead head is frozen in liquid nitrogen. In 2007 Alcor had only seventy-five fully refrigerated customers (and that included heads-only), whilst the Cryonics Institute had eighty-three.[188]

Frozen flesh might last forever, but it is still dead flesh. Cryonics is betting on a dream – even though Ettinger reckoned that the odds were 'excitingly favorable'. There is no doubt that the service offered by Alcor is very high-tech; it keeps its technology updated, and its metal 'dewars' certainly look *very* twenty-first century.[189] But, as yet, medical science has not answered the problem of how to thaw successfully live cryonic specimens. And it still has not found a way of stopping or reversing that troublesome, complicated thing called ageing – let alone a cure for death. As we shall see in the final chapter, those bodies are destined to remain in cold storage for a few years more . . .

CHAPTER 6

This Immortal Coil

Eventually, the process of aging, which is unlikely to be simple, should be understandable. In fact, in the next century, we shall have to tackle the question of the preferred form of death.

Francis Crick, *The Humanist* (1986)

Oxford, the summer of 2006   Daydreaming at my desk, I'm playing the six-degrees-of-separation game – not sideways, but backwards. I once had tea with Luke Gertler, whose father Mark, the painter, was at art school with Paul Nash. Nash knew William Blake Richmond, who was godson of *the* William Blake, the visionary artist who saw 'the World in a Grain of Sand' and held 'Eternity in an hour'. And I can almost feel the six of us, links in an intangible chain, touching hands down two centuries of time. Back each one of us goes, connected ceaselessly to previous generations. And somewhere, way down that vast procession, my ancestors walked on the plains of Africa: millions of years of natural history, natural selection, that have helped to make me exactly who I am today. It is this illimitable relationship that makes me an historian and makes me feel eternal human history at my fingertips.

Our connection with the world at such moments of imagination seems so immense, one wonders how it could ever be broken. The idea of death amongst so much life seems impossible, an unbelievable intrusion on existence. Touching Oxford's stone college walls, warm with the afternoon sun; watching a wood pigeon's flight against the clear summer sky, the fall and snap of its wings as it rises to fall again in a beautiful, seesawing arc; the earth in our garden after rain – ripe, rich, and filled with life. The possible absence of me from these enduring experiences is unthinkable.

The Suffragette Margaret Nevinson recollected her son, the future futurist painter Richard Nevinson, coming home after having seen a funeral. The little boy was 'feverish and terrified, clinging to me and crying: "Oh! mother, don't you die, I'll never let them put you in a hole, let us go on here for ever."'[1] And I remember how, as a young child, I sat on the back stairs at night, long after I was supposed to be asleep in bed, crying at the tragedy of death. I must have been only five or six years old – the age when we first become aware that our life and the lives of those we love do not go on forever. My mother consoled me – and told me that, when *I* grew up, people would live much, much longer: one hundred and twenty years, she told me then.

How improbable and cruel death seems to children. Then we put the thought out of our mind, mostly, until we grow older and feel time and age creep slowly upon us. No wonder that, through the four hundred years of this book, some visionary people have sought a way of escaping this lethal trap of growing old and wearing out. Yet it seems as though my mother might have been right: we *are* steadily living longer – and looking younger, too. In England and Wales between 1911 and 1920, an average of seventy-four people a year reached their hundredth birthday; by 1990 this figure was almost 2,000; only seven years later, it passed 3,000. This statistic is partly due to the increase in the birth rate a century before, together with the sharp fall in infant mortality and a decline in general mortality from childhood to the age of eighty, which has caused the geriatric population to grow. However, demographic analysis still demonstrates that 'the most important factor in the explosion of the centenarian population – 2 or 3 times more important than all other factors combined – has been *the decline in mortality after 80*.'[2] Even after ninety-five, mortality in developed countries has fallen, and continues to fall.[3]

Compared to Sir Francis Bacon's day, when to live to eighty was considered remarkable, this is in itself an incredible achievement. Yet the maximum lifespan has also increased. Research in Sweden has shown that between 1860 and 1889 the oldest anyone lived to be was 105. But this maximum has gradually risen over successive decades. It reached 112 by 1994, and Britons now in their thirties have a one in eight chance of living to a hundred.[4] Indeed, what is interesting about extreme long life is that the Gompertz curve, which shows mortality rates accelerating as we age, actually starts to slow down after our mid-eighties: having made it that far, it appears that only the very fittest have survived.[5] Even an 110-year old has a 45 per cent probability of living to see their next birthday.[6]

Possibly the first fully verified person to live 110 years – a 'super-centenarian' – was Katherine Plunkett. Born into an aristocratic Irish family in Kilsaran, County Louth, on 22 November 1820, she died at Ballymascanlon House, County Louth, on 14 October 1932. She was 111 years and eleven months.[7] The oldest verified person to have *ever* lived is widely recognized to be Jeanne Calment (Figure 6.2). Born in Arles, France, on 21 February 1875, to Nicolas Calment, a ships carpenter, and to his wife Marguerite, Jeanne died in Arles on 4 August 1997, aged 122 years, five months and two weeks. An official certificate confirms the date of her birth, whilst numerous authoritative national and civil registers – including census returns and marriage records – clearly map and date the course of her life. This has all been carefully researched and authenticated.[8]

It is now over a decade since Jeanne Calment's death, and her record still stands. There's no chance of it being broken soon, either. In the summer of 2007 there was no one in the world alive over the age of 114.[9] Was she simply an anomaly – what the Darwinist Edwin Ray Lankester considered a freak of nature, a 'monstrous and abnormal phenomenon'? Or is it just a matter of time before her record is inevitably broken?

If it is, a woman will almost certainly break it: the top ten of the world's oldest living humans form a woman-only list (with US and Japanese nationals predominating).[10] In fact, the oldest living *man*, Tomoji Tanabe of Japan, who was born on 18 September 1895, comes in at 29th on the Gerontology Research Group's regularly updated on-line table.[11] It has long been recognized that women, on average, live longer, and that the longest-lived humans are female (the opposite of what both Aristotle and Sir Francis Bacon had supposed). A graph plotting increasing female life expectancy at birth in the different record-holding countries shows that it has risen at the remarkably steady rate of three months a year for the last 160 years.[12] It shows that any woman born today in the record-holding country (currently Japan) can expect to live *on average* a year longer than a woman born there four years earlier. If this line is projected far enough into the future, eventually even the *average* female age at death is going to surpass 122 years: it should do so around the year 2140.

Although *average* life expectancy at birth is steadily increasing, *maximum* lifespan does not appear recently to have increased very much. Nonetheless, a leading biogerontologist, Professor Tom Kirkwood of the Institute for Ageing and Health at Newcastle University, England, suggests that 'there is no reason to believe that Madame Calment's longevity is the limit of human survival'.[13] Rudi Westendorp, Professor of Gerontology and Geriatrics at Leiden University Medical Centre, agrees. Westendorp has recently written that the 'demographic data suggest that the limit of our biological design has not yet been reached'.[14]

But how are we going to reach that limit, be it 122 years, or something well beyond? For this to happen, surely some sort of medical intervention will be required? Could such an incursion into the previously impossible soon happen? Over the last two decades there has been something of a revolution in gerontological research. From once being a bit of a research backwater, ageing has become a hot topic. Change – and, with it, optimism – is definitely in the air.

Recent research shows why. In 2003 Andrzej Bartke, Director of Geriatric Medicine at Southern Illinois University, announced that, through a combination of genetic alteration and nutritional restriction, a mouse at his Aging and Longevity Research Laboratory had had its life prolonged to almost five years – the equivalent to a human being living to between 180 and 200 years. This was not simply prolonged old age, either: it was prolonged *youthfulness*. And Professor Bartke's success has been only one of a number of similar recent achievements of substantial life extension in small model organisms. Could the same soon be possible for humans?

Twenty-first-century scientists such as Dr David Gems at University College, London, who studies ageing and life extension in nematode worms, are cautiously confident. As he told me, technically speaking, it 'ought to be possible to manipulate ageing' in humans.[15] And there is a man who's more than cautiously confident; he has been called lots of things in the press: a 'maverick', a 'prophet', 'a rogue researcher', 'the man who wants to live forever' and 'the man who would murder death'. He's been described by a prestigious American professor of surgery as one of 'the most prominent proponents of anti-aging science in the world' and 'more than a man; he is a movement'.[16] A celebrity in the worlds of gerontology and of what has been dubbed 'posthumanism', he lives and works in Cambridge, England. So I went to meet him.

Dr Aubrey de Grey's beard is remarkably long and, he insists, quite free from grey hair. He has no truck with those who think alcohol might be an impediment to longevity, and a beer commonly accompanies an interview. We meet in his regular haunt, The Eagle on Bene't Street, around the corner from King's and Corpus Christi Colleges, to enjoy a couple of pints of real ale.[17] De Grey is eager to talk, to promote the work of his appropriately named Methuselah Foundation, and to explain what he calls his 'Strategies for Engineered Negligible Senescence' (SENS).

Figure 6.1 Aubrey de Grey: front cover of February 2005 edition of *Technology Review*. Recent advances in genetic engineering have seen scientists massively extend the lives of nematode worms, fruit flies and laboratory mice. Aubrey de Grey has led the fight to see such techniques applied to humans. He believes that the first person to live a thousand years may already be alive today.

Born in 1963 and educated at Harrow School, de Grey studied Computer Science at Trinity Hall, Cambridge, and worked in computer programing and artificial intelligence before turning his attention to ageing. In 2000 he was awarded a PhD in biology by Cambridge University. Intelligent, perceptive, witty and personable, de Grey has helped place biogerontology firmly on the public map. A single question is enough to set him off. He talks enthusiastically about his work and ideas, before pausing, waiting attentively for your next request for him to share his knowledge and insight – or to see if you might dare suggest he could be wrong. When we met in July 2006, he was still employed by the Department of Genetics at Cambridge University, but has since started working full-time for the Methuselah Foundation. He also edits the journal *Rejuvenation Research*, has founded the Institute of Biomedical Gerontology and runs SENS conferences, which bring together leading researchers from around the world. And he has recently published a book, with Michael Rae, titled *Ending Aging: The Rejuvenation Breakthroughs that Could Reverse Aging in Our Lifetime* (2007).[18] The man's energy is almost boundless.

De Grey is unequivocal. Reversing ageing is about saving lives. He asserts that Louis Pasteur, through his germ theory of disease, can 'be credited without much argument as the person who has extended more lives by more years than anyone in history'. Indeed, he reckons that Pasteur 'may well have added on the order of a billion person-years to human life'.[19] It is an inspiring statistic, and it is clear that de Grey hopes to do much the same. As he is keen to point out, around 100,000 people every year die. Prolong their lives by just twenty-four hours, and cumulatively that's 274 life years saved at a single stroke. But de Grey reckons on prolonging our lives by much more than just a few hours: he is looking at years – hundreds of years.

Whilst he does not believe that an 'anti-ageing medicine worthy of the name' yet exists – and is fairly certain one won't for at least another fifteen to twenty years (perhaps the reason why he has arranged to have his head cryonically frozen in case of premature death) – de Grey confidently suggests that the first person to live a thousand years may well be alive today. They may already be in their sixties.[20] 'I think I'll either die of old age between the ages of 90 and 110,' he told a journalist in 2006, 'or I'll die from an accident at an age so great that it cannot be predicted.'[21]

These are staggering claims, and it is obvious why Dr de Grey has upset some scientists in his field. To understand how he thinks this will be achieved, we need to start by looking at how science and comprehension of ageing have progressed almost exponentially in the past two to three decades. And, to do this, we have to understand a little more about cells – because it is at the cellular (and, by inference, at the genetic) level that all the major action is happening in ageing. It is a long and intriguing journey.

§

Understanding of cell biology has advanced enormously since Alexis Carrel's work early in the twentieth century. It is now known that the human body is made up of around sixty trillion cells – 10,000 times the current total human population of the earth.[22] Yet, as tiny as each living cell is, it too is a highly complex thing, composed of thousands of lifeless molecules that are themselves made up of individual atoms. A cell consists of numerous parts carrying out different important functions – indeed, it is a veritable chemical micro-laboratory of protein production and of minute replicative, destructive and reconstructive activity. When Sir Francis Bacon contemplated the strands of life and death in the early 1600s, the microscope had only recently been invented, and a cell was the little room where a monk lived in a monastery. Today, with electron microscopes, we can penetrate to the very essence of our molecular being.

Two key parts of the cell are the lysosome and mitochondria. The lysosome is the cell's defence mechanism, a membrane sac containing powerful enzymes that can digest and break down unwanted materials which penetrate the cell wall. Mitochondria are the cell's engine, absorbing fuel molecules and combining them with oxygen to release energy.[23] At the core of each cell (except red blood cells) is the nucleus. The nucleus contains chromosomes, of which there are twenty-three pairs in the normal human cell nucleus. Each chromosome is composed of two long strands of deoxyribonucleic acid (DNA – which is also contained in the cell's mitochondria, where it is known as mitochondrial or mt DNA). A lot of this DNA, which is the code for life, appears to be random and meaningless. As Dr John T. Potts and Professor William B. Schwartz put it:

Upon initial inspection, DNA resembles an undecipherable encyclopaedia written in an obscure language consisting of just 4 seemingly
randomly repeating letters [A for adenine, C for cytosine, G for
guanine and T for thymine]. Much of this encyclopaedia appears to
be meaningless. Sandwiched among this gibberish are the important
parts, the genes, which compose only 1–1.5 per cent of the entire
sequence of nucleotides. The genes are a blueprint for all the
components of the body.[24]

Each of the twenty-three pairs of human chromosomes carries
hundreds or even thousands of these genes, which determine all the
functions of our bodies – and also (as we shall see) some of their
malfunctions. The Human Genome Project, completed in 2000, has
identified each of these genes – perhaps around 20,000, which are
involved in making proteins.[25] But mapping the genome is only the
start of the story of understanding our bodies and their processes – and
science is nowhere near the end. Having a map of our genes is one
thing; interpreting and using it is something else entirely.

Obviously, genes are hugely important in understanding the processes
of life and death. It was the Cambridge University scientists James
Watson and Francis Crick who in 1953 famously discovered that the
DNA, from which genes are formed, is not a single chain: it is a *double*
chain, linked – as they realized – in a helix formation. It is this double
helix that gives DNA the crucial property of being able to divide, copy
and thus *replicate* itself. But in terms of our current understanding of
how and why we age and die, genes are not quite as important (for the
time being) as the larger entity of which they form a part: the cell.

Remarkably, the description I have just given of the make-up of a
human cell is also accurate for virtually *any* single-celled organism. As
the American science writer Boyce Rensberger explains, 'one of the
most profound insights to emerge from modern cell biology is that all
cells, even those from species as different as a yeast and a human, carry
out the same fundamental housekeeping chores of life in exactly the
same way'. The individual human brain cell is thus 'not more
complicated than a one-celled creature inhabiting pond scum'.
Though their differing special functions give them distinctive shapes
and appearances, the fundamental processes for sustaining life in *all*
cells are ultimately identical.[26]

So, when we think about an individual human cell, we need to think of it as if it were like *any* single-celled micro-organism. That is, as if it were an individual living thing, breathing, feeding, excreting, moving, responding, reproducing in its own right. Because that is actually what it *is*. The human body is what Rensberger vividly describes as 'a community of organisms, a huge colony of extraordinarily selfless citizens, each forsaking independent existence for the good of the colony'. Our body is 'a republic of cells, a society of discrete living beings who have, for the good of the society as a whole, sacrificed their individual freedoms'.[27] As the German biologist Matthias Schleiden rightly speculated in the nineteenth century, each cell in any multi-cellular organism leads a 'double life' – its own individual existence, and that of the organism of which it forms a part.[28]

Of course, most human cells don't have much of an autonomous existence: they are packed and bound tightly together as part of our liver, our brain, our toes. But some do. White blood cells, for example, function semi-independently, patrolling ceaselessly for germs and bacteria: when they catch intruders, they almost literally eat them. Other 'natural killer cells' search for kin that have turned cancerous: when found, they exude a substance that destroys the mutant cell.[29] (It is defence mechanisms such as these, together with the antibodies produced by our immune systems, which destroyed the monkey glands Voronoff grafted onto his human patients.) Some specialist cells repair and rebuild bone; others heal wounds. Sperm cells have the most obvious independent existence: each possesses a tiny tail to power itself along in search of its female equivalent, the ovum.

The potential independent existence of *all* living cells can be seen in what is now a simple laboratory process, in which cells are separated from their companion tissue and enabled to divide and grow independently. In the right conditions, as Rensberger explains, individual human cells 'will revert to a way of life much like that of their evolutionary ancestors . . . crawling about their culture dishes like amoebas on a pond bottom, feeding on nutrients in the water, reproducing by cell division'. If kept alive long enough, certain cells will even 'remember' their original roles: skin cells will start growing and form into layers of membrane; cells from the female breast will manufacture and secrete milk protein; heart muscle cells can form into large fibres that may even beat rhythmically. Brain cells can also be induced to multiply in a petri dish, linking to form synapses and behaving, Rensberger writes, 'as if they were assembling themselves into a primitive brain'.[30]

It all sounds like something Victor Frankenstein would have dreamt of – though it also indicates how replacement body parts (such as for a skin graft, or, one day, for grafts of whole organs) could be grown outside the body. Unlike the cells of an amoeba, however, these detached, normal human cells do not have the potential to continue growing forever *in vitro*. Eventually they *will* stop dividing, and they *will* die. (At least, this is true most of the time. In 1996, Dr Lynn Allen-Hoffmann, Professor of Pathology at the University of Wisconsin Medical School, was studying ways of slowing down the ageing process in skin cells when she accidentally discovered a mutated strain of non-cancerous, 'immortal' human cells. In 2001 she founded the company Stratatech to develop medical uses for this patented cell line. She named it NIKS: Near-diploid Immortalized Keratinocyte Skin.)[31]

This is the exact opposite of what Alexis Carrel claimed for his 'immortal cell', which seemingly survived, endlessly dividing in the laboratory, from 1912 to 1946. It was not until 1961 that two American cell biologists, Leonard Hayflick and Paul Moorhead, were able to prove that Carrel's experiment was flawed and – fundamentally – plainly *wrong*.[32] In what was at the time a controversial article, Hayflick and Moorhead described their experiments, which showed that, rather than being immortal, normal human cells cultured *in vitro* go through a process of accelerated ageing involving a fairly predictable number of cell divisions before replication ceases.[33] That is, the replicative period of normal cells from large multi-cellular organisms is *finite* – and the number of divisions can be counted.[34]

It also appeared that the number of those divisions was directly related to potential lifespan. In the case of humans, this number is of forty to sixty divisions. For mice, which live at most for three years, it is between fourteen and twenty-eight divisions; for the giant tortoises of Galapagos, which can live as long as 175 years, it is ninety to 120 divisions.[35] Lifespan seemed to be relative to these seemingly impassable cell division barriers, now known as 'the Hayflick limit'.[36]

Leonard Hayflick, who has been credited with launching the field of cellular gerontology, also showed that this limit was not affected by 'temporal' time. If cells taken from large, multi-cellular organisms are frozen and subsequently defrosted, they will 'remember' how 'old' they are, and how many times they had divided before freezing. Hayflick has kept cultured human cells frozen in liquid nitrogen for over thirty years; on thawing, they have continued to replicate – though only the number of times they would have done had they never been frozen.[37] Likewise, cells taken from an old person will divide fewer times *in vitro* than those from a younger subject. It has also been found that sufferers from the genetic disorder Werner's syndrome, who age at something like twice the rate of normal humans, have a lower Hayflick limit.[38] Cell division, it would appear, is ultimately the underlying cause of senescence and death.

So why does a human cell eventually lose its ability to continue replicating? In 1975 a student at Hayflick's laboratory, Woodring Wright, working in the wake of suggestions made by A. M. Olovnikov and James Watson, showed that the 'counter' or 'replicometer' setting the Hayflick limit was located in the DNA of the cell nucleus. Each time a chromosome divides and replicates, a small sequence of DNA at the end of each strand is lost: it's a bit like knotting two ends of a piece of rope, and then cutting the knot in order to retie it to a new length of rope. Each time, a small length is lost. The finite ends of these pieces of DNA – which have also been likened to the plastic tips that prevent the ends of shoelaces unravelling – are known as telomeres. It is their shortening that ultimately stops cells from further replication.

Eventually, like the end of a rope or a shoelace which is too frayed and ragged at its ends to be tied together, the telomeres become so short that the enzymes which copy the DNA for cell division no longer work properly. When this happens, it signals danger for the cell, and it stops dividing.[39] Cells that reach replicative senescence *in vitro* do not die immediately, or even soon – if their culture medium is replaced periodically, they can stay alive for months, or even years.[40] But death ultimately occurs, and for this reason telomeres have sometimes been dubbed our 'internal clocks'.

Cell death is not the same as the death of the body, however. (Nor is it the same as ageing, as we shall see shortly.) Cells divide at different rates: brain, nerve and muscle cells usually do not divide at all once fully formed, whilst those in the hard-working gut divide quickly and last only a few days. Many cells are not completely senescent – even in centenarians.[41] In fact, for reasons we shall come to, most people die long before reaching the Hayflick limit: even the cells from an eighty or ninety year old can undergo about twenty further rounds of division *in vitro*.[42] Hayflick himself believes that the natural limit of human lifespan is about 115 years, whilst Boyce Rensberger suggests that if we 'could prevent the things that kill us sooner, our cells may have the biological potential to carry us into our mid-100s'.[43]

An important recent discovery is that the destruction of telomeres is not inevitable. An enzyme called telomerase synthesizes and elongates telomeres, adding sequences to the ends of chromosomes at each cell division.[44] This makes it possible for cells to replicate continually – to become, in fact, effectively immortal. But this is not necessarily a good thing – at least, not yet.

There are already two sorts of 'immortal' cells in humans. Both sorts can die – and most do – but they also have the potential to live on for ever. The first sort, known as germ cells, are found in our specialized reproductive organs: they are sperm and ova, and the sex cells located in the gonads that produce them. Unlike the soma cells making up the rest of the body, germ cells have only one instead of two strands of our twenty-three chromosomes. When these male and female 'immortal' germ cells unite, their DNA strands combine to create a new being: our offspring. It is these immortal cells that link us illimitably to our primaeval African ancestors. And they link us, too, to those *very first* living cells that appeared on earth, somehow, somewhere, billions of years ago. All life, in this respect, *is* an endless, immortal chain.

The other immortal cells occur in tumours. They are well known to us all as cancer cells. Continually dividing, if left unchecked they will eventually kill us – and, left without a host to feed them, they too will then die. If removed from their host, however, cancer cells – unlike Alexis Carrel's chicken cells, or the normal human skin and brain cells described above – will live and divide without limit *in vitro* – forever.

In 1951, a young African–American woman named Henrietta Lacks was diagnosed with cervical cancer, and cells from her vigorous tumour were sent to a researcher at Johns Hopkins University. Radiation treatment failed, and Henrietta Lacks died. But cells cultivated from her original tumour are still alive in laboratories around the world – they have even been sent into space.[45] Known in medical parlance as HeLa, they have been described as possibly 'the first "immortal" human cell line'.[46] Henrietta Lacks, of course, is not alive; but what was once part of her still is.

Understandably, anti-cancer research has focused a lot of attention on telomerase. The US-based biopharmaceutical company Geron Corporation, for example, is developing cancer treatments targeting telomerase inhibition. It aims to develop other drugs that will activate telomerase in tissues impacted by senescence, injury or degenerative disease, enhancing cell repair and function. We will return to this shortly, when we look more closely at how modern science and medicine might soon be capable of vastly prolonging human life.

§

Though the Hayflick limit is clearly important in understanding how long we live, it does not answer the question of *why* we age. As Hayflick has written, 'Virtually all biological events from conception to maturity seem to have a purpose, but aging does not. It is not obvious why aging should occur.'[47] So, far from having any fixed answer, the question 'why?' remains vexing. But it needs to be explored if we are to understand how it might be overcome.

Why we age can be examined from two perspectives: firstly, at the cellular level; secondly, at a species level – through the lens of evolution by natural selection. Ageing is much trickier to answer from the first perspective than from the second. In fact, as Rensberger observes, 'it is not clear why a cell that has ceased dividing needs to die. Whether it stopped dividing within the body or reached its Hayflick limit in a laboratory flask, nobody really knows what happens inside a cell to bring death.'[48] As we have seen, death does not *have* to be a necessary part of life: unicellular organisms can go on forever. Even certain larger organisms do not appear to age. The sea anemone has been found to continue reproducing whilst showing no obvious signs of ageing for a period of over eighty years.[49] The hydra – a tiny freshwater organism less than an inch long – is also able endlessly to regenerate new selves asexually without showing signs of ageing: this is a sort of auto-cloning that renders it effectively immortal. And there are pine trees in California and Tasmania that live for thousands of years.

Yet there is, it appears, something particular to most multicellular organisms that makes them, eventually, die. As Rensberger explains, it appears that when single cells 'banded together hundreds of millions of years ago to form multicelled organisms, each cell had to subordinate its freedom to the good of the community, to the good of the republic. Each cell also surrendered its immortality.' The price of combination 'may have been the gradual destruction of any given cell's ability to renew its parts indefinitely'.[50] In a sense, therefore, Alexis Carrel was right: senescence *doesn't* happen at the cellular level at all.

Well, in fact, for us humans it clearly does. Normal human cells in culture are seen to undergo hundreds of functional changes as they age and approach their Hayflick limit. Almost all aspects of their biology, chemistry and behaviour are affected by these events, some types of cells being affected more than others.[51] There is a number of theories explaining why our cells age and lead to those processes we associate most closely with senescence, such as failing eyesight, hearing and memory, loss of muscle strength and stature, and greying hair. In all likelihood it is a complex combination and interaction of factors – together, perhaps, with some that currently remain unknown to us – that explain why our cells (and, with them, our organs and bodies) age.

There are numerous probable causes. All are related to the gradual build-up of microscopic damage and/or waste accumulation at the cellular/molecular level. And all point to processes reflecting modern definitions of ageing: namely, that it is a process of cumulative degenerative change leading to the increasing likelihood of death (the statistical process demonstrated in the Gompertz curve). None will kill us straight off, and none are 'deliberate' genetic mechanisms: rather, they are the highly complex, random and undesirable side effects of life in a multicellular organism. As Kirkwood explains, 'eventually they build up in such numbers that they overwhelm the body's capacity to keep its life support systems running', and we die.[52]

It is certainly highly unlikely, therefore, that there is a *single* cause of ageing; it is, rather, what is commonly referred to as a 'mosaic' of mechanisms. It is perhaps surprising that, well over two millennia after Aristotle pondered the problem and four centuries after Sir Francis Bacon posited some possible answers, we still do not fully understand the processes of senescence and death.

Free radicals currently top the list of theories of ageing damage. Also known as Reactive Oxygen Species (ROS), they originate inside cells, either as the side effect of normal mitochondrial metabolism, for example when certain compounds react (especially) with oxygen, or when radiation (including heat and light) or certain toxins break cell molecules apart. Free radicals are thus atoms or molecules no longer rooted (from the Latin: *radicalis*, 'having roots') in the larger molecule to which they once belonged. Unbound from their parent, they have a free electron – which seeks to bind itself to another electron, splitting off more molecules, and causing further damage – or producing toxins, or mutant, cancerous cells. It is the free radicals produced by oxidation that cause 'ageing' to most objects: in materials containing iron, they reveal themselves as rust; and they are what turns food rancid.[53] Their action is suspended at very cold temperatures – and hence freezing keeps food (as well as corpses) 'fresh'.

Our cells are exposed to an endless daily bombardment of free radicals.[54] Healthy cells can cope with the loss of a few molecules, and possess defence mechanisms that sweep up destructive rogue atoms. Dangerously damaged cells can even commit suicide, a process known as 'apoptosis'. But, like rainwater slowly and invisibly penetrating a house, gradual toxin and free radical damage can escalate, or they can outrun the body's natural defence mechanisms, or they can simply become too expensive in terms of energy-use to make defence or repair worthwhile. As we grow older, this damage increases as our cells' ability to preserve and restore themselves deteriorates. Brick by brick, the old house with its ageing tenant crumbles, and ageing itself becomes an increasing risk factor of susceptibility to many diseases, such as cancer.[55]

The action of free radicals can also cause cellular mutation (something that may also occur from mistakes in cell division). Replication of DNA is a phenomenally complicated business, involving millions of chemical transactions at the molecular level. Occasionally, large mistakes – mutations – occur. And as we age, smaller mistakes, known as epimutations, also become more common. A tiny error occurring early on in the DNA of a cell nucleus, or in mitochondrial DNA, is replicated and can accumulate. Some mutations may result in cancers; others may inhibit the body's ability to fight cancers or other forms of damage.[56] As Tom Kirkwood and Caleb Finch (another leading contemporary researcher on ageing) explain:

There is now much evidence in mammals for the more-or-less progressive accumulation throughout life of somatic DNA mutations, epigenetic defects . . . and abnormalities of chromosomal number and structure. Together, these lesions result in a general drift away from the unique genomic identity of the zygote. The age-related increases of certain cancers, e.g. of the breast and prostate, are associated with somatic cell gene mutations.[57]

Added to the problems of free radicals and mutation is the gradual harm caused by waste accumulation. As well as being part of the whole bodily republic, each cell is an independent organism in its own right, 'breathing', feeding and excreting. Sometimes a cell feeds on substances it cannot fully digest. This is not necessarily a problem for a single cell organism, which can simply excrete such waste – in the same way that the human stomach and intestines will expel bad food by vomiting or diarrhoea. But at the multicellular level this is not so easy. As our vomit includes acidic digestive enzymes from our stomach, so the same happens at a microscopic level. This acidic excretion would injure neighbouring cells: thus junk accumulates inside the cell, which is unable to excrete its waste.

Like a blocked drain in a house, slowly filling it with rancid waste, the cell gradually loses its ability to function properly.[58] This process may explain the decline in the secretion of hormones that is displayed by most glands in the endocrine system as we age.[59] Atherosclerosis, or the hardening of the arteries, is another manifestation of this waste accumulation – a gradual process that can lead to coronary heart disease, stroke or cardiac arrest.

Even when they are not dividing, all cells, including those of the muscle, brain and nerves, are continually leading an inner life of their own. This includes their continued renewal. Old parts are broken down, new ones assembled: over a lifetime, most of a cell's molecules and structures other than the DNA will have been remade thousands of times. This process results in the gradual accumulation of brown, indigestible debris known as lipofuscin. Lipofuscin accumulation can lead to functional decline and disease, with cells gradually losing their ability to function properly. The 'republic of the body', as Rensberger aptly puts it, 'gradually disintegrate[s]'.[60] Lipofuscin accumulation is a particular problem in the brain, and is one of the most consistent anatomical features of human ageing. Its effects can also be seen in specific problems such as macular degeneration in the eyes. But it appears to have an all-round effect of gradually reducing operative cellular capability.[61]

Further age damage occurs through the loss of functional cells over time without their subsequent replacement. The declining capability of many organs is often 'largely attributable' to the gradual loss of cells and tissue, 'even in the absence of overt disease'.[62] Like bricks, tiles, paint and plaster falling from a house without being replaced, cell loss without replacement is an underlying cause of decay: neuro-degeneration, osteoarthritis, renal failure, liver dysfunction, muscle loss and declining bone mass all result in the common ailments, complaints and physical decline characteristic of old age. An opposite problem is the accumulation of unwanted cells – such as inactive immune cells and fat cells. Also, fat cells can be lost where we need them most – under the skin, for example – and appear in places where they are unwanted. And, on top of all that, there is the constant pull of gravity.

Life is a battle against all these forces. When you add them up, the outcome – degeneration, decay and death – does not appear all that surprising.

§

As these processes make clear, the better care we take of our bodies, the longer we may resist the cellular effects of ageing. Everyday experience shows that some people age 'better' (or more slowly) than others. Nor is ageing an even process: our different parts age at different rates. Some effects of ageing, such as hair loss, are merely aesthetic: a man who goes bald at thirty will not die younger than one who retains his hair into his eighties. It just makes him *look* older. Our age in years, therefore, does not necessarily correlate directly with what gerontologists sometimes call our 'biological age'.

Exposure to sunlight is a clear cause of ageing. Cells of the skin are necessarily among the fastest replicators, for the skin is our buffer against the world; it takes constant, high-level free radical damage and needs regular replacement. Not surprisingly, skin cancers are among the most common. Exposure to high doses of radiation is one of the most obviously malign influences. It can quickly cause mutation at the cellular level, which leads to cancers and, eventually (and sometimes rapidly), death.

Cancers of the digestive system are another common form. Again, our guts do a lot of work, and we are, quite literally, what we eat. Food is vitally important both in reducing cellular damage and in enhancing repair. Scientists are increasingly discovering that natural chemicals found in foods such as soya beans, broccoli, cabbage, cauliflower and fatty fish (for instance salmon and mackerel), as well as in many fruits and nuts, aid the body's natural ability to repair damaged DNA. Antioxidants, in particular vitamins C and E, are thought to be especially valuable, whilst foods high in cholesterol (which contributes to atherosclerosis) are bad. The traditional diets of Japan and the Mediterranean region have been identified as especially beneficial for long life. Newspapers thus regularly hail the latest wonder foods: 'Miracle diet halts ageing', to cite an erroneous 2006 front-page banner headline in one leading British tabloid newspaper.[63] The market for so-called 'functional foods' – ones with 'enhanced' ingredients such as omega 3, or supposed cholesterol-reducing margarines, or 'probiotic' yoghurts – is worth billions of pounds a year, and is growing rapidly.[64]

Every so often a new drug or hormone briefly emerges as the latest medical cure. In the late 1990s, melatonin prompted books such as Marvin Cetron and Owen Davies's hugely optimistic *Cheating Death: The Promise and the Future Impact of Trying to Live Forever* (1998). Produced by the pineal gland, melatonin controls many of our life cycles: research by Dr Walter Pierpaoli suggested that pineal transplants into young mice doubled their natural lifespan. Cetron and Davies (science writers both, not physicians) spoke confidently of 'something approaching immortality' and of a 'transition to a post-mortal world' within the next fifteen or twenty years.[65] That was ten years ago. Melatonin is now best known as a way of overcoming jet lag. There are likely to be many more such false dawns.

As Alexander Graham Bell showed, our genetic stock – our parentage – plays a part in how long we live. Recent studies of mice, fruit flies, nematodes and human identical twins have led to the conclusion that 'Genetic estimates for the heritability of life span . . . range from 10 per cent to 35 per cent.' The remainder of the variance 'is divided into gene interactions . . . and environment'.[66] So we need to eat well and to get plenty of exercise. We need to avoid environmental toxins, both in our food and in the air – including noxious pollutants such as tobacco, exhaust fumes and other man-made poisons that spoil our atmosphere, as well as the plastics, metals and pesticides that impregnate our food and the soil in which it grows. Look too closely at the man-made world, and it is easy to be disgusted at the careless way we treat it and ourselves.

The 'six non-naturals' of Hippocratic medical theory are as valid today as they were in ancient Greece – more so, in fact, as we actually know so much more about how our bodies and the environment interact. We need to look out for beneficial foods and vitamins: Roger Bacon and Tobias Whitaker's recommendation that we drink regular quantities of red wine has since proved good advice. The focus on the mind has also proven correct: regular mental exercise, such as cross-words and other sorts of puzzles, reading and continuing to learn and to challenge our brains, all help in holding back ageing – even Alzheimer's.[67]

If we took greater care in our food, exercise, air, work, leisure, environment, mental health, many and more of us would clearly have longer, healthier lives. Certain interventions are already able to improve the quality of life as we age – such as hormone replacement therapy, organ transplants, hip replacements, and various therapies for cancer. These help us age more easily, and some definitely save lives. Yet they do not prolong them beyond the current maximum. Indeed, as Leonard Hayflick has pointed out, even if cancer and cardiovascular disease were cured tomorrow, that would offer only a marginal extension to average life expectancy.[68] Even if a cure was found for *all* diseases that currently appear on death certificates, most people, he suggests, would still die of *something* by the age of about a hundred. 'These centenarians', Hayflick explains, 'would still not be immortal. . . . They would simply become weaker and weaker until death occurred.' We would then, he points out, have to 'invent new terms to write on death certificates, or return to using that old term, "natural causes", attributing death to the inexorable normal losses in physiological function that are the hallmark of aging'.[69]

Yet, as we get richer and lazier and expect medicine to cure our ills rather than take responsibility for our own health, things look like they may be getting worse. As well as helping to slow cellular damage, food can obviously also be harmful: in particular, the build up of fats and salts that can contribute to cardiovascular diseases. Stress and hyper-tension have also been found to contribute to early death. In western countries, the most common cause of death after the age of sixty-five is not cancer or Alzheimer's: it is atherosclerosis, the thickening and hardening of the walls of the arteries. Beginning early in life, and influenced partly by our lifestyle choices, in later life it may result in heart attack, angina or stroke.[70] Many western countries are even facing what has been called a crisis of obesity: the number of grossly over-weight adults in the UK has almost trebled in the past three decades.[71] Our increasingly sedentary lives, supplemented by diets of processed foods high in salt, fats and chemical additives, are leading to extreme weight problems and increasing risk of the late onset of diabetes. It has been predicted that the next generation may actually see a *fall* in average life expectancy.[72] The complaints of 'luxury' and bad diet, which have resounded since the sixteenth century, have never rung so true as they do today.

Are we now to fall back into the lap of luxury against which George Cheyne railed? In this respect, much of the advice laid down over the last half millennium by Cornaro, Bacon and Cheyne on how we might live longer, healthier, fitter lives remains absolutely valid – but is still all too often ignored.

Yet restricted diet has been proven both to lower the incidence of disease *and* to prolong life in many living organisms. In the 1960s and 1970s, Morris Ross of the Institute for Cancer Research in Philadelphia developed some interesting experiments undertaken in the 1930s at Cornell University by Professor Clive McCay. Between them, McCay and Ross proved that rats fed on a nutritious but reduced calorie diet could live up to 60 per cent longer than rats allowed to feed freely.[73] One of the most vociferous promoters of using caloric restriction for prolonging *human* existence was Roy Walford. Born in San Diego, California in 1924, Walford had an early and intense fascination with ageing. Even while he was a schoolboy, it struck him that our allotted span was 'simply too short to permit a satisfying exploration of the world's outer wonders and the realms of inner experience'. We were, he complained, 'shelved at the mere beginning of our understanding'.[74] Walford, therefore, considered it the 'task of the modern researcher into aging' not simply to 'steal the fruit from the Tree of Life', but 'to make off with the tree itself . . . and to plant it in our midst'. Failure to do this, he felt, would mark the end of medical progress.[75]

Walford qualified as a doctor at the University of Chicago in 1948, and in 1954 took up a position at the School of Medicine at the

University of California, Los Angeles. In the late 1960s Walford began his own experiments in caloric restriction (or CR; also known in slightly different contexts as dietary restriction or DR). In 1983, in his book *Maximum Life Span*, Walford laid out the history and theory of 'dietary restriction': that is, a diet low in calories but still high on all the necessary vitamins and minerals – what has been called 'under-nutrition without malnutrition'. In 1988 he co-authored a book with Dr Richard Weindruch of the National Institute on Aging at Bethesda, Maryland. In *The Retardation of Aging and Disease by Dietary Restriction* they declared that 'dietary restriction' had 'emerged over competitor methodologies as the only procedure that can actually and strikingly retard aging in homeothermic vertebrates'.[76]

The ability of CR both to slow ageing (keeping laboratory specimens younger for longer) and significantly to prolong life has been proven in rats, mice, nematode worms and fruit flies. It has not yet been shown scientifically to work in humans, though there have been suggestive one-off cases. There was Luigi Cornaro – dubbed by Walford 'our pioneer friend' – who lived nearly a hundred years, back in the sixteenth century, on a tiny diet. And there was Dr Alexandre Guénoit, president of the Paris Medical Academy and author in 1931 of *How to Live a Hundred Years, or The Art of Prolonging your Days*. Guénoit addressed various techniques for a long and healthy life, but in particular recommended a light diet with moderate amounts of wine. His daily intake was about 1,500 calories, and he actually died (in 1935) aged 102.[77] But, as Walford readily admitted, two isolated cases do not make a scientific sample.

The strongest evidence Walford had were studies of the inhabitants of the Japanese island of Okinawa. Accurate legal birth records have been kept in Japan since 1872, and research indicates 'that the incidence of centenarians on Okinawa is 2 to 40 times that of any other Japanese island'. Other government studies have shown that the total caloric intake of school children in Okinawa is only 62 per cent of the 'recommended intake' for Japan. Compared to the average Japanese diet (which is, of course, already a good one for life expectancy), Okinawans ate plenty of fish and vegetables, but much less sugar. Deaths from cerebral vascular diseases, malignancy and heart disease were all well under *half* the average for the rest of Japan. And deaths per 100,000 inhabitants in the 60–64 age-bracket, over the decade 1977–87, were

1,280 in Okinawa, compared to 2,181 elsewhere in Japan.[78] These were remarkable statistics – and it needs to be noted again that the Japanese population is itself one of the most long-lived in the world.

Walford was highly optimistic of the prolongation of life that could be achieved by CR: up to 130 to 140 years, he reckoned in 1983, if the 'very stringent measures' the programme demanded were adhered to. He followed this programme himself, and for two days a week ate nothing at all, with a healthily 'supplemented' diet on the remaining days. In his last book on this subject, *Beyond the 120 Year Diet*, published in 2000, he was even more optimistic. He predicted that this method might prolong life 'up to perhaps 150 to 160 years, depending on when you start and how thoroughly you hold to it'.[79]

For Walford, CR was not an end in itself. It was a method to live longer in order to live long enough to reap the benefits of what he called 'future breakthrough possibilities' in anti-ageing technology. As he stated in 1983, he believed that by the end of the twentieth century 'advances in fundamental biology will allow a substantial slowing down or even halting of the aging process'.[80] Of course, these have still not happened. I watched *Space 1999* as a child in the 1970s and thought how far away and how exciting the twenty-first century looked. Almost anything would be possible then. Here we are, and there *are* all sorts of technological and medical advances neither I nor Walford could ever have imagined. Yet science had still not caught up with his predictions.

Walford was not a lone practitioner, however. He co-founded the Calorie Restriction Society in California in 1994, and its website promotes his ideas.[81] One member breakfasts on a single egg white, whilst steamed vegetables, berries and salad leaves all come highly recommended. Another member, a forty-eight-year-old Californian, explained his motivation in simple terms in 2006: 'I don't want to get decrepit and die,' he said. 'I decided that there is nothing more important than life, so I should do what I could to prolong it.'[82]

Early results from studies undertaken by Luigi Fontana of Washington University since 2002 on about thirty practising subjects suggest that CR 'is highly effective in reducing the effects of arteriosclerosis in humans'.[83] So it waits to be seen if any of the society's members will ever exceed Jeanne Calment's 122-year maximum recorded lifespan. Walford himself, however, died in 2004, a few weeks short of his

Figure 6.2 Jeanne Calment: photograph. A few days before her 122nd birthday, Jeanne Calment posed for press photographers. Five months and two weeks later, on 4 August 1997, the world's oldest woman died in her home town of Arles, France. She is in the only person verified to have lived into their thirteenth decade.

eightieth birthday – from motor neurone disease – still awaiting those 'future breakthrough possibilities'.

§

Walford and Weindruch believed that dietary restriction worked somehow at the cellular and glandular level. Their studies of calorie restriction in mice suggested that in such circumstances the DNA's capacity to repair itself declined more slowly than normal.[84] A common side effect of CR is reduced libido/fertility, and in the case of rodents it has been suggested that CR 'increases lifespan by reducing metabolic rate' – what has been called the 'rate-of-living effect'.[85] These processes may have an evolutionary purpose: in times of dearth, the optimum life-drive is simply to survive long enough until times of plenty return – a sort of process of enforced hibernation. It is almost as if the body temporarily shuts down its energy-expensive functions, such as reproduction, whilst focusing instead on protecting its core operations, such as repair.

Current research indicates that diet affects lifespan through a number of different mechanisms. Over the past two decades there have been some remarkable laboratory experiments with various model organisms which have led to substantial prolongation of life. In the 1980s, Michael Klass and Thomas Johnson of the University of Colorado isolated long-lived mutant *Caenorhabditis elegans* nematode worms. The mutation was eventually located in a single gene, which they named age-1.[86] In the early 1990s, Dr Cynthia Kenyon at the University of California, San Francisco, discovered a mutant nematode with double the normal twenty-day lifespan. This was seemingly caused by a defect in a single protein-manufacturing gene known as daf-2: all such mutant nematodes enjoyed both youthfulness *and* greater longevity. The mutation affected the production of an insulin-like hormone, which in turn influenced metabolism and growth. By 2004, through a combination of genetics, environment and micro-surgery, researchers had achieved up to a six-fold increase in nematode lifespan – the equivalent in human terms, as Kenyon points out, would be 'healthy, active 500-year-olds'.[87]

Researchers such as Professor Linda Partridge at University College, London, have extended fruit fly lifespans by up to 50 per cent, whilst 70 per cent increases have been achieved in mice. Genetic engineering has also boosted lifespan in yeast, nematode worms and fruit flies. It is possible that a repair enzyme, named *Sir2*, is produced in times of environmental stress – for example, because of lack of food – which links these cases of life prolongation neatly into the dietary restriction debate.[88] The fact that this repair process is present throughout the animal world – including in humans – helps biogerontologists like Aubrey de Grey to make their startling predictions, because what is possible for small organisms should also hold true for large ones like us.

The clear suggestion from the latest research is that diet and genetics play complex roles in extending lifespan, perhaps through raising the levels of production of antioxidant enzymes and/or through increasing the uptake of intestinal nutrients (via insulin signalling).[89] Another interesting line of research suggests that, in certain cases of dietary restriction in nematode worms and fruit flies, life extension may partly be caused by the prevention of 'bacterial invasion', by reducing their noxious effects on elderly specimens. Posing the question: 'Is food toxic?',[90] this line of enquiry may in part vindicate some of the arguments made by Metchnikoff early in the twentieth century. The hope is that, in the future, techniques – either through genetic engineering (which would manipulate our genes *in utero*, or earlier) or even through the oral administration of detoxification drugs or genetic inducers – will replicate these effects without the need to indulge in the disheartening process of calorie restriction.

The prolongation of life along these lines will not simply be a case of making old people live longer. It will be about making us age much more slowly. It is certainly one of the most exciting current developments in gerontology. In an effort to promote awareness and research in exactly this field, de Grey's Methuselah Foundation has established the Methuselah Mouse Prize. Its fund already tops $4 million, with an award being paid out from it whenever a research team beats the standing record for the longest-lived lab mouse. It's an exciting proposition – and one that invites everyday people around the world to help to contribute to advancing a scientific project of monumental ambition.[91]

§

Studies of calorie restriction suggest that its life prolonging effect has an evolutionary origin. This requires at least a brief examination, in order to address the question of why some animals live so much longer than others. Is it really possible that *humans* could live substantially longer, or is our 120 years a maximum a *nec plus ultra* beyond which it is simply foolish to aspire? Humans already do very well in the longevity stakes, living much longer than any other primate – in the region of 50 per cent longer than chimpanzees and gorillas. In fact, humans are the longest-lived land-based mammals.[92] Yet we don't live as long as a Galapagos tortoise, or a bowhead whale – or a sequoia pine tree. Clearly it is *possible* for cells functioning in large organisms to live well beyond Thomas Parr's supposed 152 years. And, if humans have already evolved to live twice as long as their closest genetic cousins, should it not be physically possible to live longer still?

Tom Kirkwood is well known for his theory explaining why we age, which famously came to him whilst taking a bath; it was first published in an article in *Nature* in 1977, and it developed from ideas formulated in the 1940s and 1950s by the biologist J. B. S. Haldane (1892–1964), the zoologist Peter Medawar (1915–87) and the evolutionary biologist George Williams.[93] Kirkwood was responding to two long-standing biological suppositions: that ageing is *inevitable* (because, like any machine, we simply have to wear out eventually) and *necessary* (because we are genetically programmed to die in order to prevent overcrowding and to make room for our descendants). Kirkwood argued, however, that ageing is neither inevitable *nor* necessary – the subtitle, in fact, of his 1999 book, *Time of Our Lives*.[94]

We have seen how in large organisms such as humans the germ cells (those sex cells, including sperm and ova, located in the gonads) are effectively immortal. When the male and female germ cells unite they start to divide, and within a few days a small bundle of cells known as a blastocyst forms. The blastocyst consists of stem cells: these are very important, as they have the potential to become *any other cell or tissue* in the body. As the embryo grows, stem cells will develop into all the limbs, organs and other parts of the body. Almost all of those cells will be what is known as soma cells. A small number will be the 'immortal' germ line: the sperm and the ova. Obviously, most of these 'immortal' cells will never survive. A woman is born with roughly a million eggs, but by the end of her menopause none remains. Only if she has children will her 'immortal' germ line have survived to carry itself on. Likewise for a man: he will have produced billions of sperm in a lifetime, but few (if any) will live on.

So how does all this explain ageing? The answer is evolution, and what Kirkwood calls the 'disposable soma' theory. What matters to any species' ultimate survival is its germ line; everything else – the somatic cells – are simply building blocks: they are there to carry and advance the germ line. If it is true that people are frequently preoccupied with sex, then there's a very good reason for it: sex is our *raison d'être*; we are literally built around our gonads. Brains take second place in the human race. We are the walking, talking tools of our germ lines. *They* are immortal; *we* are simply their sophisticated – but nevertheless entirely disposable – carrying cases. Like the driver of a complex hire car that's replaced every eighty years or so, the germ line drives forever onwards. We individuals, meanwhile, are left to decay, discarded by the roadside as our germ line disappears happily into the sunset in the form of our children, and our children's children, and our children's children's children, and so on.

The reason for this disposability, Kirkwood argues, is that making *any* new cells is a tricky, energy intensive process, in which mistakes are very occasionally made. As he realized in his bath back in 1977, to be successful, the germ line has to be highly accurate. This is not quite so important, however, for the building-block somatic cells: 'Sooner of later the soma is going to die by accident,' he explains. 'Might it not be better to save energy,' ensure the near perfection of the germ line, 'and make somatic cells in a more economical way, even if this results in them ageing?'[95] For the same reason, why expend more energy than is necessary on maintaining somatic cells if, in the evolutionary battle that is the natural world, they're eventually going to be killed anyway – either by external accident or by predators? Professor Michael Rose found that the 'Methuselah flies' whose lives he had substantially prolonged in his laboratory by artificial selection did not survive as long in the wild as their shorter-lived, unmanipulated relatives. They had 'sacrificed hardiness for longevity, an exceedingly high price to pay'.[96]

Kirkwood illustrates his evolutionary theory with an imaginary animal he calls a mobbit – a sort of cross between a mouse and a rabbit. Mobbits are immortal in so far as they never age – but they can be killed, either by accident or by predators. (This must happen eventually: as Professor Steven Austad has calculated, if humans ever did achieve immortality, their average life expectancy would be 1,200 years. This is how long he reckons it would take statistically before an accident would eventually and inevitably kill them.)[97] In Kirkwood's model, the average annual chance of a mobbit dying by accident or predation is 50 per cent. Thus, although a five-year-old mobbit is physiologically *identical* to a one-year-old, there will be far more one-year-old than five-year-old mobbits running around. In fact, statistically speaking, with an annual mortality rate of 50 per cent, only one in thirty-two mobbits will live to age five. With the number of older mobbits halving each year, although no mobbit *ever* dies of disease or old age, only one in a million will live to be twenty!

Given this steady, merciless rate of attrition, where is the point in expending energy on maintaining and repairing somatic cells, and on preventing ageing? As Kirkwood explains, suppose a mistake occurred in the DNA replication of one particular mobbit. In Kirkwood's mobbit model, a mutation suddenly means that one particular mobbit can expend less energy on somatic maintenance and repair. This allows it to expend more energy on growth and reproduction instead. This genetic error comes at a price, however: this mobbit is not immortal any more. It will age and die by the time it is twenty.

But what does that matter? There is only a one in a million chance it will live to twenty, anyway! It simply does not *need* to live that long to fulfil its reproductive function. Unlike its immortal brothers and sisters, this particular mutant, mortal mobbit will reproduce more – and stronger – offspring. In the tough world that the mobbit inhabits, mortality is actually an *advantage*. Very gradually, its fitter, mortal offspring will expand within the mobbit population. 'The mutation will spread,' as Kirkwood explains, 'and ageing will enter the population.'[98] Immortality will be rejected.

The wide distribution of senescence as a phenomenon in nature suggests that ageing is much older than *homo sapiens*, and a process akin to the one just described actually began millions of years ago. Kirkwood's theory may also partly explain why the sea anemone does not appear to age. With few natural predators, little chance of accidental death, and no long-lived, non-dividing cells (such as a brain), it can invest plenty of energy in somatic defences and repair. The degree of maintenance required to live even a few days is enough to live forever.[99] Immortality, therefore, is *theoretically* possible: it is just that the human race – not as individuals, but as a species – is better off without it.

§

Ageing and the debilities that come with it are an evolutionary phenomenon.[100] We deteriorate simply because we are not built to last; physical immortality is not inherent within the human frame. Yes, we need to live long enough to reproduce, but life *beyond* reproductive age is a bonus.

Natural and sexual selection work together to weed out the useful from the useless. That is why the young – generally speaking – have fit, strong, physically attractive bodies that quickly repair themselves when damaged, are more resistant to disease, and will attract a mate. If they die too young, *ergo*, they don't reproduce: any flaw or weakness is not passed on to their children, because they do not *have* any children. Gradually, that flaw will be weeded from the gene pool. Those healthy, clever, quick, resilient and/or attractive enough both to survive and to mate will pass on those useful traits through their germ line. This process has repeated itself over millions and millions of years.

After reproduction, our continued survival becomes increasingly unimportant. The diseases and debilities of older age – Alzheimer's, cancer, arteriosclerosis, heart disease, pneumonia – have little evolutionary impact; they are not weeded out, because the people who die from them have already reproduced their fatally flawed germ line. Indeed, some of those very things that may have worked in our favour during our early life may, in old age, prove the very causes of disorder or decay. Genes that help regulate cell growth early in life may later prove responsible for cancer.[101] We are like drag-racing cars, well built for a fast race over a short distance. If a key component in our physical engine gets us across the finish line faster and fitter (with sex and reproduction being the finish line), it does not matter much if it breaks down or malfunctions further along the track: it has already successfully served its purpose.[102]

Of course, our ageing ancestors would not have been worrying about the debilitating diseases facing them in their old age: most of them never got there. As their strength and senses waned as they grew older, they became increasingly vulnerable to predation, and of decreasing value to their social group. As very old males don't reproduce much, and because of the menopause old human females don't reproduce at all, those genetic factors that helped them live long were not *specifically* passed on. Long life is not an asset in the jungle or on the savannah. But reproducing young – and often – is. Now, as we increasingly live life outside the jungle, we are living longer and discovering more and more of those flaws in our genetic make-up that strike us only in older age. As Sir Peter Medawar put it, ageing is revealed 'only by the most unnatural experiment of prolonging an animal's life by sheltering it from the hazards of its ordinary existence'.[103]

It does appear, however, that there is some evolutionary impetus for humans to live longer. Longevity is, in part, heritable. The fact that humans long outlive all other primates has been attributed to the evolution of our more advanced and larger brains, increasing our intelligence and encouraging sociable support networks.[104] These things protract and postpone the age at which we reproduce. Older, post-reproductive adults do have a use in society: they possess wisdom and knowledge, and can assist in the caring of the young. The menopause, which does not occur in any other primate, may also be evolutionary: women stop having babies in order to help in the care for the vulnerable children they already have, and for their own grandchildren.[105]

These are positive trade-offs, which all encourage longevity. If a child survives partly because of the intervention of one of its long-lived grandparents, it will have inherited some of that ancestor's genetic make-up – an instance in which old age *is* an inheritable asset. As humans (in the wealthy West at least) reproduce later and later in life, it is likely that we will continue to evolve to live longer: Michael Rose proved that this happens with the 'Methuselah flies' he bred at Sussex University in the late 1970s. But it will take a long, long time to see the results in humans: it's reckoned that the doubling of human lifespan compared to that of our ape-like ancestors took between two and three million years.[106]

It is necessary, therefore, that we look at quicker routes to prolonging life.

§

'If a genetic shortcut to immortality exists,' Francis Fukuyama has observed, 'the race is already on within the biotech industry to find it.'[107] It is through genetic engineering and allied research that a cure for death is envisioned.

The completion of the Human Genome Project has laid out a map that might lead the way, and this is being diligently explored. It is increasingly clear that some of the damage affecting us in later life happens very, very early on in our existence – what we might call a design *fault*, rather than a design *constraint*. Chromosome 8, for example, carries over 1,400 genes, among which is the one that, if defective, causes Werner's syndrome. Having unusually short telomeres, individuals with Werner's syndrome age and 'grow old' more rapidly. This – like other, more acute forms of premature ageing – indicates how human ageing is not simply about time passing, but is very much the result of cellular/genetic processes that could eventually be 'reprogrammed' and thus overcome. Other disorders linked to single-gene defects, such as Down's syndrome and cystic fibrosis, can be identified at a very early, embryonic stage – and this 'unideal' future life can be aborted even before it is fully formed. This is widely considered ethical. Genetic engineering already offers the possibility of embryo screening – to prevent certain diseases, or to predict certain specific characteristics, for instance height, hair colour, intelligence. Such traits could be chosen – though this is currently considered unethical. Could longevity be one of them?

Other genetic disorders do not become apparent until much later in life. Whilst Parkinson's has seemingly been mapped to a single gene (in chromosome 4), Alzheimer's is a far more complex neurodegenerative disorder, associated with a number of genetic defects across a number of chromosomes.[108] Nonetheless, the possibilities of emerging technology are enormous, and Fukuyama wonders if in the future death 'may come to be seen not as a natural and inevitable aspect of life, but a preventable evil like polio or the measles'.[109]

As anyone will know who followed the final years of the paraplegic actor Christopher Reeve (famous, so ironically, for his role as Superman), stem cells will play an important role in helping to repair physical damage. This will include the damaging effects of ageing – that is, if some of the ethical issues precluding their use (particularly in the USA) can be overcome. Development came too slowly to save Reeve's life, but it surely won't be too long before surgery and genetic engineering, based on stem cell research, will be able to restore damaged nervous systems.

Stem cells are 'harvested' from embryos left over from *in vitro* fertilization treatment. They were first isolated and grown in culture in 1998, and are important because they can self-replicate endlessly, giving rise to every type of cell in the body. But the removal of stem cells kills the embryo of which they are a part: hence the ethical conundrum. The promised advances from stem cell therapy include introducing these cells into the body to replace and repair damaged or aged organs, such as the heart or liver.[110] It is already possible to grow human tissue and organs outside of the body.

Advances made in this direction include the successful cloning, in 1996, of an adult mammalian cell, 'Dolly' the sheep. Given enough time, money, willing surrogate mothers and (more importantly) legal sanction, it should be quite possible to do the same with humans. Some people have seen human cloning as a way of 'immortalizing' the self. But a clone is no more 'you' than an identical twin is its own brother or sister. A cloned 'you', however, would be able to provide the 'real' you with a stock of spare organs, marrow, blood cells or whatever was required to prolong your life. That cloned 'you' need never be more than tissues created *in vivo*, and it would never have an independent existence. It would be a sort of store for spare parts, each exactly created (or even improved) to fit you and no one else.

In the collection of essays *Coping with Methuselah: The Impact of Molecular Biology on Medicine and Society* (2004), Nicholas Wade suggested that, if we could find a way of taking a patient's own stem cells and using them to repair or replace damaged organs, '[w]e would have fashioned an extraordinary technique for overcoming many of the degenerative diseases of old age. Such an advance would lead to a quantum leap in life expectancy.'[111] Wade adds that, if we could understand how the body is able continually to replace itself and why this process eventually ceases, given what he calls 'the underlying robustness of cells and organs, it seems quite possible that human lifetimes of great length could in principle be supported once we understand how to manipulate the genes that influence longevity'.[112] And, as the biomedical ethicist Professor John Harris has observed:

In the long term it may be possible to 'switch off' the ageing process and also to maintain a repair programme in cells . . . If all the cells of an individual had their ageing programme switched off, and were programmed to regenerate, then the immortality that this would confer on the cells would be passed on as the cells multiplied and differentiated, eventually affecting every cell in the body as it was formed. The resulting children would be truly immortal.[113]

The Hayflick limit does not appear to be an impediment to this possibility. It has been discovered that *normal* human cells in culture can exceed their maximum lifespan more than fivefold by the introduction of the telomerase catalytic protein component. This discovery, as Jerry Shay and Woodring Wright have pointed out, 'has profound theoretical and practical implications that include the immortalization of normal human cells for the production of commercially important proteins'.[114] Recent experiments introducing telomerase into healthy animal cells resulted in their continuing to divide for 'over 300 generations past the time they normally would stop dividing'. The cells, furthermore, grew normally, and gave rise 'to normal cells with the normal number of chromosomes'.[115]

Shay and Wright explain that the 'eventual goal' of this research 'is to be able to take a patient's own cells, expand them in a laboratory environment, genetically engineer them to correct a particular defect, and then reintroduce them into the patient in a form that permits the cells to function in tissue-specific manner'.[116] What they call the 'immortalization' of specific cell types obviously has profound implications. As individual organs age or become diseased, it may soon be possible to grow replacements in the laboratory. This is not far-fetched. Just think of what Bacon or Descartes would have made of a human heart and lung transplant – surgical interventions almost impossible even to imagine in the seventeenth century.

§

Leonard Hayflick is very sceptical that human immortality might be possible – or desirable. In 1996 he wrote that the fraction of the US health budget invested in understanding the causes of senescence was tiny, and stated that he did not believe we had 'sufficient understanding of either the aging process or the determinants of life span to expect to significantly manipulate either during our lifetime'.[117]

Though he firmly believes we should attempt to eliminate the present leading causes of death, Hayflick sees 'no value to society or to the individual in seeking to slow or stop the aging process or to achieve immortality'.[118] In fact, he points out (rightly, in my opinion) that, if nothing is done soon to control overpopulation, earth 'will not be a place on which it is worth spending more time'. With this looming Malthusian threat, the best future scenario Hayflick envisages is 'one in which all humans reach the maximum life span, still in possession of full mental and physical abilities, with death occurring quickly as we approach, say, our 115th birthday. This is the goal that we are tacitly pursuing at this time. Virtually all biomedical research has the implicit goal of eliminating disease in all of its forms.'[119]

But it may be too late to take this stance. As David Gems told me, 'we're stuck with life extension', because – like it or not – that is where current medical research into the cure of disease is leading us. It's 'an ethical trap taking us towards the posthuman', he admits.[120] In 2005

Michael Rose, now Professor of Evolutionary Biology at the University of California, Irvine, wrote that, having spent some three decades postponing ageing in fruit flies, '[o]ne of the ubiquitous dreams of mankind, the conquest of aging, is now foreseeable. It hasn't arrived yet, but a juggernaut has begun to roll forward, and it is picking up speed.'[121] Lucian Boia states that in 2001 Rose declared his belief 'that there are already immortal people'.[122]

The prospect of considerable life extension happening soon is strong enough for John Harris to have written in the journal *Science* in 2000 that it is 'unlikely that we can stop the progression to increased life-spans, and even "immortality" . . . We should start thinking now about how we can live decently and creatively with the prospect of such lives.'[123] Notwithstanding all the other fields of medical research (or, indeed, the implementation of knowledge we already have), that could – and need – to be financed, Harris does not believe the pursuit of immortality is unethical. This is partly because he does not consider it a goal that is necessarily being pursued in its own right. Rather, 'immortality' will be a side effect of all the other medical research that is going on into curing diseases and genetic disorders.

When I discussed his ideas with Harris, I asked him why he was so interested in immortality. His response was simple: 'Because I don't want to die.'[124] This is where Aubrey de Grey steps back into the picture. He does not accept the pessimism of scientists such as Hayflick, and approaches head on the natural human fear of perpetual oblivion. Scientists such as Hayflick fail, de Grey believes, to recognize all the possible and what he suggests are 'technically feasible inter-ventions that, if applied jointly, seem likely to constitute real anti-ageing medicine'. He reckons it will take around another decade to prefect these 'interventions' in laboratory mice, and two more to translate them into humans.[125]

These interventions obviously promise far more than the latest, spurious anti-ageing creams advertised in glossy magazines. Yet they still only have the potential to add a relatively small amount of extra life: 10, 20 or maybe 30 per cent, perhaps. So when de Grey suggests that the first person to live a thousand years may be alive right now, he is not suggesting that a 'cure' for ageing is just around the corner. It may not even be as close as thirty years away. What he *is* suggesting is that, within the next few decades, we are likely to discover medical interventions that either will slow down the speed at which we age, or will result in physical rejuvenation, and will prolong our lives another few years. Then, within those years of extra life, he believes scientists will find further interventions. These in their turn will give us, or will retrieve, additional youthful years. And so the process will continue, like Zeno's arrow: through a series of infinitely small increments, human life never quite reaches its final destination – death.

As the first human rejuvenating methods kick in, de Grey argues that the clamour to provide them will almost guarantee their availability and improvement: not everyone gets cancer, Parkinson's or AIDS; but everyone grows old. We all have a personal interest in finding (and funding) a cure. This won't be prolonged old age, either; it should be a prolonged middle age – or youth, even. And we won't be faced with the question, 'Do I actually *want* to live a thousand years?' De Grey suggests that it will, rather, be a case of 'do I want to live for another ten, or twenty years?' It will be immortality in tiny steps.

De Grey is enthusiastically pointing the way forward. A theorist and activist rather than a laboratory scientist, he has adopted what he calls 'an engineer's approach' to ageing, based on his Strategies for Engineered Negligible Senescence. SENS identifies seven causes of ageing at the genetic or cellular level, all of which, he asserts, are 'potentially fixable by technology that already exists or is in active development'. SENS depends on the periodic repair of the damage that, as we have seen, occurs all the time at the cellular level, and on preventing it from 'snowballing out of control'.[126] His theoretical interventions thus include solutions to the problem of junk accumulation in cells: this, he suggests, may be removed by introducing extra enzymes to cells, enabling them to 'degrade' waste material. Cell depletion, which causes so many age-related effects, may be treated by introducing growth factors to stimulate cell division, or by introducing cloned stem cells to replace those lost.[127]

De Grey recognizes that the hardest aspect of overcoming ageing is the problem posed by cancer (which appears in more than half of all autopsies of those over eighty-five, even when it is not the actual cause of death).[128] He advocates an ambitious solution, focusing on the total elimination of the genes for telomerase – that is, of the enzyme that synthesizes and elongates telomeres, and enables cancer cells to continue replicating indefinitely. De Grey's suggestion concentrates on gene, not drug, therapy. He foresees the 'repopulation of all our stem cell populations with new ones whose telomeres had been restored *ex vivo*'. De Grey believes this rejuvenating intervention, which is somewhat akin to a bone marrow transplant and would be repeated every ten years or so, will 'maintain the relevant tissues indefinitely while preventing any cancer from reaching a life-threatening stage'. These processes, he notes, are 'already close' to being technically feasible in mice.[129] Before too long, he argues, they will be capable of implemention in humans, too.

Inevitably, some scientists are highly critical of de Grey. Though admitting to be 'intrigued' by him, Sherwin Nuland, Professor of Surgery at Yale School of Medicine and author of *How We Die* (1993), is sceptical of this desire to prolong life beyond its 'natural' term (a concept that could soon be of debatable meaning – if it isn't already). In a profile published in *Technology Review* in 2005, Nuland observed that many of de Grey's strategies are, at best, speculative. He concluded with the observation that de Grey

> has issued the ultimate challenge, I believe, to our entire concept of the meaning of humanness.
>
> Paradoxically, his clarion call to action is the message neither of a madman nor a bad man, but of a brilliant, beneficent man of goodwill, who wants only for civilization to fulfil the highest hopes he has for its future.

Nuland, like Hayflick, Fukuyama and many other traditionalists, sees death as a necessary part of life, a part of what makes us human. He considers it 'a good thing' that, in his opinion, de Grey's 'grand design will almost certainly not succeed. Were it otherwise, he would surely destroy us in attempting to preserve us.'[130] In November 2005, twenty-eight scientists, including Tom Kirkwood, signed an article criticizing de Grey and the SENS programme. They declared that it 'falls into the realms of fantasy', and dissociated themselves from him and those who were 'impressed by de Grey's ideas in their present state'.[131]

His detractors have not blunted de Grey's spirit. He seems to have an answer for every criticism that could be directed either at his theory or at the suggestion that it would be awful if humans did discover a way of immortalizing themselves. He has no desire for children, and, like William Godwin, believes that 'people acting in their own self-interest and choosing not to have kids' will solve the overpopulation threat. Anyway, he says, our priority, our 'duty', both to ourselves and to future generations, is to develop rejuvenation therapies 'as fast as possible'. We can worry about the implications later.[132]

De Grey remains confident of his prognosis for human ageing – and there are plenty who agree with him. The 'transhumanist' movement has embraced him wholeheartedly. Its prophets – men such as Ray Kurzweil and Marvin Minsky – have predicted a near future in which bioscience and nanotechnology will enhance, repair and massively advance human potential. Kurzweil and Terry Grossman have promised a *Fantastic Voyage: Live Long Enough to Live Forever* (2004). (It is notable, though, that their methods include careful diet, meditation and vitamin supplement pills – just the type of techniques that we have seen reiterated throughout this book.) Gregory Stock, meanwhile, has written about what he calls the 'metaman' and the creation of a 'global superorganism' by merging humans with machines.[133] It has even been suggested by the likes of Kurzweil that one day it will be possible to 'upload' our individual patterns of personality and memory into machines, and to continue our existence in computerized form: this, though, is a thing quite distinct from prolonging life in its fleshly form – and one that (if it ever happens) would raise all sorts of philosophical questions and dilemmas about what, exactly, it means to be human. Here, perhaps, we re-enter the territory of the immortal soul – a region I have marked as beyond the intellectual domain of my enquiry.

The Immortality Institute in the USA has funded forward-looking publications such as *The Scientific Conquest of Death: Essays on Infinite Lifespans* (2004). De Grey, who contributed to this collection, taps into the transhumanists' expectant excitement, and has attracted big money to the Methuselah Foundation – including (in 2006) a promise of up to $3.5 million from the young American internet entrepreneur Peter Thiel. The Methuselah Foundation is investing in active programmes of research, and is already funding PhD students in promising areas of biogerontology.

The British journalist Bryan Appleyard has colourfully catalogued and quizzed the transhumanist movement in his recent book, *How to*

Live Forever or Die Trying: On the New Immortality (2007). Artificial Intelligence and supercomputers, the forthcoming 'singularity' (an almost messianic event in which the rate of knowledge reaches a cumulative point of totality) and the supposed inevitability of scientific progress beyond our wildest dreams, all add up to a not-too-distant future in which immortality seems almost inevitable. It is the overspill, perhaps, of a sort of *fin de siècle* optimism, like the one that drove Condorcet and Godwin to such visionary extremes.

§

After four hundred years, it does appear that we are really getting somewhere in the war on ageing. Finally, we seem on the brink of a breakthrough. Is immortality just around the corner? It looks possible, and the scientists and philosophers I have spoken to have made me overcome some of the natural scepticism I feel as an historian: too many people over the course of this book have been too confident that they have found a route to prolong life radically. Each new medical discovery or advancement has suggested to someone, somewhere, that here is a way to cheat death: chemical elixirs, the power of the mind over matter, the control of bacteria, eugenics, hormone injections, gland grafts. Some of these may have added a little to average human life expectancy at birth; but none has ever done more than that. Why will now be any different? These things appear to come in waves of optimism: right now, we are definitely at a peak of interest in the possibility of immortality, as the recent number of books on the subject (including this one) shows.

Is this optimism misplaced? Our faith in the power of modern medicine is such that, as Stuart Spicker of the University of Connecticut's Health Center observed as far back as 1990, today in the West we have a 'tacit image of invulnerability, if not an image of full-fledged immortality'. Death is something we expect to be deferred by medicine somewhere in the undetermined future. Death is elsewhere. It doesn't, can't, and won't happen to us.[134] Even some people within the medical fraternity, Spicker observed, shared this dream. Was it, he wondered, because as a profession 'we have failed to acknowledge the intrinsic fallibility of empirical knowledge, especially medical knowledge, whose phenomenal growth since 1870 is unquestionable'?

Have we, he wondered, 'failed to appreciate the indisputable fact that all empirical, biomedical, and clinical knowledge (though the outcome of humanity's continuous quest for certainty) is intrinsically provisional, probabilistic, incomplete, and subject to reinterpretation if not downright refutation by means of experimental methods or verification'?[135]

Spicker felt – and there is truth in this observation – that medicine had supplanted religion in the Western worldview. Life, not death, now dominates our thought. He warned his fellow professionals:

> We must, in short, remain vigilant and guard against an imperious arrogance: that the profession of medicine may be tempted to substitute itself for religious authority and once again proffer the false promise of quasi-immortal life . . .[136]

That warning was given almost twenty years ago. I feel it still rings true: everything within this book has suggested that it is correct. Is it the same for genetics – or is this the final peak on the long climb upwards from the foothills Sir Francis Bacon and René Descartes surveyed so many centuries ago? Is science about to arrive at the summit, plant its flag of dominion, and declare that death itself is deceased? Or do we stand near the tip of another false crest?

I do, sincerely, believe that death *ought* to be conquerable: but this is a philosopher's belief, not a physician's. It is notable that it has mostly (though by no means exclusively) been philosophers over the past four centuries who have proposed the real possibility of human immortality.

Is it *desirable*? That's quite another question. Swift, Malthus and Hayflick have all disparaged the idea. De Grey believes that, once the technology is in place, everyone will be clamouring for it. Then, he suggests, governments will be forced to respond: if they don't, they simply won't get elected. Life extension will become the single issue of the day. Where this would leave the world – a world in which some, or everyone, might live for a thousand years or more – I leave it for others to try and answer. Though I think it *ought* to be possible, I do not believe that it *will* ever happen. Whilst I might wish it for myself, I would not wish it on the world. For if the world *is* alive – as the Gaia theory suggests, or as was thought by some, back in the seventeenth century – it has a cancer, and that cancer, sadly, is the human race:

destructive of the organism upon which it depends, and growing uncontrollably. Would not prodigiously long-lived humans make this scenario worse?

Certainly we are inching closer to this miracle of science: fear of old age and death will always drive at least some research in this subject. Bacon and Descartes both believed that this dream would be achieved in the future, once medicine had been radically transformed. And here we are, *in* the future. Medicine *has* been radically transformed, far beyond anything Bacon, Descartes or Condorcet could ever possibly have imagined. The French Cartesian La Mettrie called 'Man a machine': we may no longer think so fully of ourselves as machines, but we do plug our bodies *into* machines, and machines into our bodies: iron lungs, pacemakers, ventilators, incubators, computers – even nano-computers soon. I have titanium plates and screws holding my ankle together; my brother was cured of cancer, using chemotherapy drugs whose ingredients included platinum (which would have pleased the alchemists); a pacemaker the size of an i-pod maintains my father's heart in steady motion. It can, in fact, sometimes prove difficult to die in a hospital; court orders are sought to turn life-support machines off, or to keep them running. It is in the brain that death is now (still somewhat uncertainly) identified: with machines that can keep the heart beating and the lungs inflating indefinitely, the twenty-first century hospitalized body often only dies when we let it.

Yet none of this, as I have said already, has made any one of us live and breathe a minute longer than 122 years. Of course, there is still a long way to go yet. I will not reach Jeanne Calment's remarkable age until the last years of this century. By that time, who can imagine what will be possible medically and technologically if we continue on our current course – that is, if we do not destroy the earth along the way.[137]

'No-one gets out of here alive', sang Jim Morrison. But maybe – *maybe* – now, at last, in the twenty-first century, there's a real chance. Science and medicine seem finally to be catching up with our dreams – in ways that we could barely have imagined even twenty years ago. I am still only in my thirties; I must have a chance.

Notes

Preface

1. Jonathan Swift, *Thoughts on Various Subjects* (London, 1706).
2. Harris (2000), 59.

Chapter 1   The History of Life and Death

1. See Bacon's deathbed letter to the Earl of Arundel reproduced in Bacon (1874), 550.
2. Apart from the abovementioned letter, this account is based on Thomas Hobbes's description of Bacon's death recorded by the antiquary John Aubrey (in Aubrey (1898), Vol. 1, 75–6), and on William Rawley's 'The life of the honourable author' (in Rawley (1661), unpaginated). Is this truly what happened? Who can say for certain: no famous death goes unembellished or unromanticized. For an alternative account, suggesting that Bacon accidentally poisoned himself with opiates and nitre, see Jardine and Stewart (1999), 502–11.
3. See Bacon (1857–74), 2nd edn (1876–83), Vol. 3, Part I, 222. Still the best book for the context of this whole intellectual debate is Webster (1975).
4. Aubrey (1898), 1.71.
5. Alexander Pope (1734), *The Essay on Man*, Epistle 4, 3. 296.
6. *Oxford Dictionary of National Biography* (Oxford: Oxford University Press, 2004), hereafter *ODNB*.
7. Bacon (1638), 'To the present and future ages'.
8. Two translations of the book were published in 1638: I have used the second, made by Bacon's former secretary, William Rawley.
9. Gale (1971), 391–9.
10. Rawley (1661), 'The life of the honourable author', unpaginated.
11. Samuel Sorbière (1709), 32.
12. *Frier Bacon his Discovery of the Miracles of Arts, Nature and Magick* (London, 1659), 30.
13. Overton (1644), 1.
14. Steele (1688), 21.

15. Almond (1999), 42–3.
16. Quoted in Kassell (2001), 358.
17. Both quotations are in Almond (1999), 45, 42.
18. Genesis 3: 22–3.
19. Ibid.
20. Quoted in Kassell (2001), 360.
21. Talbor (1672), 2.
22. Faber (1677), 12.
23. Rare counter-arguments to this orthodox view were William Rabisha's radical reversal of the Fall, *Adam Unveiled and Seen with Open Face* (London, 1649), and Isaac La Peyrère's *Prae-Adamitae* (London, 1655).
24. See Egerton (1966), 576–7.
25. See Gruman (1966), 21.
26. Basse (1619), 9–10.
27. Browne (1646), 299–300, 343–4; Genesis 5: 21. Augustine had made this point in the *City of God*: see Egerton (1966), 576.
28. Maynwaringe (1670). Genesis 6: 3 records, in the time before the Flood, that 'his days shall be an hundred and twenty years'.
29. Psalm 90: 9–10. The Bible is not a book consistent in its information, and various lengths of human lifespan are proffered.
30. Boyd (1629), 231.
31. Thomas Hobbes, *Leviathan*, Part I, chapter 13.
32. See Dobson (1997), 236–55.
33. Grassby (2001), 121.
34. See Smith (1976), 126.
35. Smith (1982), 193.
36. Ibid., 194; Evelyn (1955), Vol. 5, 518–19.
37. Grassby (2001), 122. See Laslett (1991), table 6.3. Hanserd Knoylls: 1598–1691; Christopher Wren: 1632–1723.
38. Hynes (1995); see P. W. Hasler, *The Commons, 1558–1603* (1981), Vol. 1. 383.
39. See *ODNB* for biographies of both men.
40. Bacon (1638), 134–5, 159. According to André Du Laurens, eighty was the natural terminus of human life; see his *Discourse of the Preservation of the Sight of Melancholike Diseases; of Rheumes, and of Old Age* (London, 1599), 172–7.
41. Bacon (1638), 244, 241. Bacon's source for Katherine Fitzgerald, Countess of Desmond, was probably Walter Raleigh's *History of the World* (1614). It seems likely that the countess was at least in her nineties when she died in 1604. See Thoms (1873), 95–104.
42. Temple (1701), 112.
43. Abbot (1996), 133. In 1538 all parishes in England were charged with the obligation to keep registers of baptisms, marriages and burials, but it is likely that births – like other significant life moments – were dated by other major events, such as wars, floods, coronations, and the like. In less literate cultures, this practice has remained widespread until relatively recently.
44. Hart (1633), 7–8.
45. *The Obituary of Richard Smyth*, Camden Society 1st series, no. 44 (1849), 77, 87; Wood (1892–4), Vol. 2, 461 and 476.

46. Peter King, *The Life of John Locke: With Extracts from his Correspondence, Journals, and Common-Place Books* (1830), 244–6, quoted in Laslett (1999), 23.

47. Abbott (1996), 136.

48. Cited in Steele (1688), 16.

49. Temple (1701), 112.

50. The Boyle Papers, Royal Society of London, MS RB/1/17/2. Steven Smith observes that 'the limited evidence does suggest that a good many people did live to experience old age' in seventeenth-century England: Smith (1976), 126.

51. Steele (1688), 17, and 'The epistle to the reader'; Inigo Jones (1573–1652), in a letter to the Earl of Pembroke, quoted in Abbott (1996), 143.

52. Thomas Fuller (1647), 261.

53. See Thomas (1976).

54. There have been lots of good studies on this subject: see for example Botelho and Thane (eds) (2001), Pelling and Smith (eds) (1991), Thane (2000).

55. See Hart (1633), 7–8, and van Helmont (1662), 810.

56. Keynes (1966), 223. The autopsy report was printed in John Betts's *De ortu et natura sanguinis* (1669), 319–25; an abstract of the report appeared in the *Philosophical Transactions* (1668), iii. 886–8. On sex and age, see McManners (1985), 80–1.

57. See Keith Thomas, 'Thomas Parr', *ODNB*.

58. See Taylor (1635): his poem is preceded by an account of how Parr was discovered and brought to London.

59. Ibid.

60. Keynes (1966), 221; Fuller (1952), 480.

61. See Harley (1994), 10–11, who points out that Harvey 'took whatever opportunities he could find to perform or observe autopsies'.

62. Keynes (1966), 224.

63. Evelyn (1661), 21.

64. Keill (1706–7), 2249.

65. Ibid., 2252.

66. Robinson (1695–7), 268.

67. Arrais (1683), 105.

68. Bacon (1640), 190.

69. Andrew Marvell, 'To his coy Mistress', in *The New Oxford Book of English Verse* (Oxford University Press, 1972), 334.

70. Maynwaringe (1670), 34.

71. Smith (1666), 7–8.

72. Ibid., 117–18.

73. Ibid., 156.

74. Wellcome Library MS 6175, fol. 99.

75. Bacon (1638), 270–1.

76. 'This my lord Bacon speakes of, but not mentioning his name, in his *Historia vitae et mortis*': Aubrey (1898), Vol. 1, 223–4.

77. Bacon (1638), 270–1.

78. Ibid., 269–70.

79. Aristotle, 'On youth and old age', ch. 4, 469[b] 18–20, in Aristotle (1910–31), Vol. 3.

80. Ibid., 469[a] 24–70[a] 20.

81. Ibid., chs 23 and 24.
82. Aristotle, 'On the length and shortness of life', ch. 5, 466^b 5–16, in Aristotle (1910–31), Vol. 3.
83. John Archer (1673), 'To the reader'.
84. Hart (1633), 4–5.
85. On the progress of this debate on whether or not old age was to be considered an illness from classical times to the early eighteenth century, see Schäfer (2002).
86. Quoted in Giglioni (2005), 136. See also Rees with Upton (1984), 64–5. Though Bacon is elsewhere a critic of Aristotle's philosophical methodology, he cites him on six occasions in the *History of Life and Death* and appears to be more indebted to Aristotle than to Galen or Avicenna, whose names go unmentioned in the essay. Hippocrates' name appears only twice (once as an example of long life, since it was said that the Greek physician had died at the age of 104).
87. McManners (1985), 80.
88. Roger Bacon (1683), 2–5.
89. Egerton (1966), 578.
90. Roger Bacon (1683), 1.
91. Roger Bacon had observed that, as 'the World waxeth old, Men grow old with it: not by reason of the Age of the World, but because of the great Increase of living Creatures, which *infect* the very Air'. Ibid., 1.
92. William Russell, *A Physical Treatise, Grounded not upon Tradition nor Phancy, but Experience* (London: printed for John Williams at the Crown in St Paul's Churchyard, 1684), 8.
93. Translator's footnote to Roger Bacon (1683), 6–7.
94. Vaughan (1633), 121: this was the seventh edition. Thomas Burnet debates this point at length, explaining why this would have been a slow rather than an immediate decline: see Burnet (1697), 130–1, 146–7.
95. Bacon (1638), 110–11.
96. See *The Posthumous Works of Dr Robert Hooke* (London, 1705), 322.
97. Arrais (1683), 6–32.
98. Ibid., 102–6.
99. First translated into English in 1634, by 1727 this handbook had reached a fourth English edition under the title *Sure and Certain Methods of Attaining a Long and Healthful Life*.
100. Cornaro (1727), 13–15, 20, 26.
101. Ibid., 94.
102. Vaughan (1633), 153, 121.
103. Whitaker (1638), 67.
104. Ibid., 70.
105. According to the *ODNB*, he may have been the Tobias Whitaker, son of Francis Whitaker, who was baptized on 1 May 1601 at St Dunstan and All Saints, Stepney, Middlesex.
106. [See] Genesis 1: 29.
107. [See] Genesis 9: 3.
108. Tryon, *Pythagoras his Mystick Philosophy Reviv'd, or, The Mystery of Dreams Unfolded* (1691), 246–7, quoted in *ODNB*.
109. John Evelyn, *Acetaria: A Discourse of Sallets* (London: B. Tooke, 1699), 137.

110. Genesis 1: 28; 4: 1. Biblical support for a degenerative theory might also be found in the third commandment, 'for I the Lord thy God am a jealous God, visiting the iniquity of the fathers upon the children unto the third and fourth generation': Exodus 20: 5.
111. Pettus (1674), 35–6.
112. Whitaker (1638), 58.
113. Ashmole (1658), 47–8: he suggests in his preface that an Englishman had written this manuscript, in around 1600.
114. See Wear (2000).
115. Bacon (1638), 344.
116. Ibid., 394–5, 183, 387–91.
117. See Walker (1972), 121. On Bacon's conception of *spiritus*, see again Walker (1972), 121–30. The 'sensible' soul found in all living things was distinct from the 'rational' soul, found only in humans, which had been imparted into Adam by the breath of God; ibid., 122–3.
118. *ODNB*.
119. In this he was influenced by the Italian philosopher Bernardino Telesio's *De rerum natura juxta propia principia* (1587) and by his follower Agostino Donio's *De natura hominis* (1581): see Rees with Upton (1984), 65–6 and Walker (1972), 121–30.
120. Bacon (1638), Preface (unpaginated), 339.
121. Ibid., Preface (unpaginated).
122. Ibid., 5.
123. Ibid., 166–7, 178.
124. Ibid., 279–80.
125. Ibid., 270–91.
126. Ibid., 220–1.
127. Ibid., 221–35.
128. Ibid., 292–7.
129. Ibid.; see Canons IX and X, 215–17.
130. Ibid., 188–99.
131. Ibid., 200–6.
132. Ibid., 240–1.
133. Ibid., 37, 241–54.
134. Ibid., 243, 247–8.
135. Ibid., 256–65.
136. Ibid., 330–5.
137. Bacon (1996), 344; in his thirtieth essay 'Of regimen of health', he observes: 'age will not be defied' (404).
138. *Novum organum* (1620), Aphorisms, 116. Bacon 'produced practically nothing' of parts 4, 5 and 6 of the *Instauration*; Rees (1975b), 171.
139. Bacon (1874), 534.
140. 'The life of the honourable author', in Rawley (1666), unpaginated. On Bacon's use of nitre as a medicine, see Jardine and Stewart (1999), 506.
141. 'Baconiana medica' (1679), 155–61.
142. Bacon (1638), 379, 425.
143. Ode 31 from the *Manes Verulamini*, published by John Haviland almost immediately after Bacon's death.
144. Bacon believed that there had been 'only three revolutions and periods of

learning . . . one among the Greeks, the second among the Romans, and the last among us, that is to say, the nations of Western Europe'. He felt persuaded 'that this third period of time will far surpass that of the Graecian and Roman learning'. Quoted in Guibbory (1986), 45, 50.

145. Daniel 12: 4; Guibbory (1986), 50–1.

146. Isaiah 65: 20. This, however, is the New American Standard Bible translation: the King James Version is far from clear in its meaning, which may explain why Bacon does not reference it.

147. Bacon (1638), 'Preface'.

148. Anne Conway to Lord Conway, 2/9 October 1651, in Nicolson (ed.) (1930), 36–7.

149. D'Argenson, 'Pensées sur la vie et sur la mort', *Journal et mémoires*, ed. E. J. B. Rathery, 9 vols (1859–67), Vol. 5, 214–15; quoted in McManners (1985), 78.

150. Bacon (1638) 1.

151. Boyle (1999b), 206.

152. Bacon (1640), 19–20.

153. Francis Bacon to Father Fulgentio, probably October 1625, in Bacon (1874), 532 (this is the editor's translation of Bacon's Latin). Fulgentio's letter directly addresses matters arising from Bacon's *History of Life and Death*.

154. Bacon adds that the inhabitants of New Atlantis possessed a '*water of Paradise*' which they made '*very Soveraign for* Health *and* Prolongation *of* Life', as well as certain 'Chambers of Health . . . *where we qualifie the* Air *as we think good and proper for the* Cure *of divers* Diseases, *and* Preservation *of* Health . . . *We have also Fair and large* Baths, *of several Mixtures, for the* Cure *of* Diseases *and the restoring of* Mans Body *from* Arefaction: *And other for the* Confirming *of it in* Strength *of* Sinews, vital Parts, *and the very* Juyce *and* Substance *of the* Body'. Bacon (1658), 27–8; see also Serjeantson (2002).

155. Bacon (1640), 67–8.

156. See Aubrey (1898), Vol. 1, 71.

157. Bacon (1640), 69–70.

158. Ibid., 64.

159. Ibid.

160. Selden, *Historie of Tithes* (1618), quoted in Guibbory (1986), 3.

161. T. Fuller, *Historie of the Holy Warre* (1639), 'the epistole Dedicatorie'.

162. Ashmole (1652), Prolegomena (unpaginated).

163. For a verification of Hobbes's dates, see *ODNB*.

164. Aubrey (1898), Vol. 1, 83, 331; Sorbière (1709), 40. See also Hobbes (1994), 196.

165. Hobbes (1994), 179.

166. Ibid., xxiii.

167. Aubrey (1898), 1.331–90.

168. Sorbière (1709), 27.

169. Aubrey (1898), 352.

170. Choron (1963), 134.

171. 'An elegie upon Mr. Thomas Hobbes of Malmesbury, lately deceased' (London, 1679). My thanks to Dr Huw Price for bringing this pamphlet to my attention.

172. Descartes to Huygens, 5 October 1637: '*Les poils blancs qui se hatent de me venir m'avertissent que je ne dois plus étudier à autre chose qu'aux moyens de les retarder. C'est mainteneant à quoi je m'occupe.*' Quoted in Lindeboom (1979), 95.
173. Quoted in Gruman (1966), 78.
174. Descartes to Huygens, 18 June 1639, quoted in Lindeboom (1979), 94.
175. Descartes to Mersenne, 9 January 1639, quoted in idem, 94.
176. René Descartes, *Oeuvres*, edited by Ch. Adam and P. Tannery (Paris: Cert, 1897–1913) (13 vols), Vol. 4, 329, quoted in Carter (1983), 7.
177. Baillet (1693), 259–70.
178. Descartes (1649), 101–2.
179. Ibid., 101–2; see Shapin (2000), 131–54.
180. Descartes (1649), 127. On Descartes' medical studies, see Carter (1983) and Lindeboom (1979).
181. Descartes (1649), 102.
182. See Gruman (1966), 79, who quotes from P. Des Maizeaux (ed. and trans.), *The Works of St Evremond, with the Life of the Author by Des Maizeaux* (2nd edn, London, 1728), ll.xli–xlii. Descartes's reference to 'la vie éternelle' can be found in British Library Add. MS 4470, fol. 4, 'An account of the first meeting between René Descartes and Sir Kenelm Digby'.
183. 'The description of the human body'; see Carter (1983), 20.
184. Hartlib, 'Ephemerides', 1650, Part 2 (February to May): The Hartlib Papers, Sheffield University, 28/1/54A–B (hereafter *HP*).
185. *Extra ordinarisse Posttijdinghe* (10 April 1650), quoted in Lindeboom (1979), 94.

Chapter 2   The Elixir of Life

1. Evelyn (1661), 12, 17, 19.
2. Henry Alford (ed.), *The Works of John Donne, Volume 3* (London: John W. Parker, 1839), 575.
3. Pepys (1972), 165, 204–5. The mortality figures and the description of London during the plague are from Pepys (1972), 165–208, 214; 201, 171.
4. The following description is based on ch. 5 of Thomson (1666), 'An historical Account of the dissection of a pestilential body; and the consequents thereof', 70–142.
5. The incubation period for bubonic plague is from as little as one and a half to as many as ten days: Thomson, therefore, would already have been infected by the bacteria *before* his autopsy of the Petticoat Lane corpse. The chances of survival after infection without modern medical intervention are roughly 50:50. Starkey and Thomson's medicines would have been useless.
6. Van Helmont (1662), 751–2.
7. See *HP*, 'Ephemerides', March 1650/51, D–D4; 'Ephemerides' 1651, Part 1, 28/2/6A; Newman (1994), 1–2, 62, 209–10.
8. George Starkey to Robert Boyle, after 19 April 1651, in Boyle (2001), Vol. 1, 100. Starkey ends one of his letters to Boyle with the salutation, 'Good health, most honoured man for now and for eternity!' Starkey to Robert Boyle, 3 January 1652, ibid., 111.
9. Newman (1994), 82.

10. Boyle (2001), Vol. 1, 156: Hartlib to Boyle, 28 February 1654. In 1662 John Ward, who had moved to London to study medicine and chemistry, wrote in his diary: 'Starky seems to bee a careless idle fellow one yt is given to tipling and spending': Wellcome Library, MS 6175, fol. 134.

11. Starkey (1658), 152.

12. Starkey (1657), 80.

13. Bacon (1640), 170.

14. See Paracelsus (1661), 35.

15. For an illuminating analysis of this changing vocabulary, see Newman and Principe (1998): their article explains how the words alchemy and chemistry were effectively interchangeable in the seventeenth century.

16. Johann Hartmann (1568–1613) was appointed to the chair of Chymiatria at the University of Marburg in 1609, the first known university appointment in chemistry. In 1683 Robert Plot was appointed the first Professor of Chemistry at Oxford University.

17. Alan Debus, 'Alchemy', in *The Dictionary of the History of Ideas: Studies of Selected Pivotal Ideas*, edited by Philip P. Wiener, 14 vols. Vol. 1, 28–34.

18. Michela Pereira writes that in the Hellenistic tradition 'iksir' was the name of an 'immortality powder'; Pereira (1998), 27.

19. *The New Oxford Dictionary of English*, edited by Judy Pearsall, Oxford: Oxford University Press, 2001, 40.

20. See Newman and Principe (2002), 1800–81.

21. Ecclesiasticus 38: 4, 7.

22. This was the belief that plants look like the parts of the body they heal: walnuts, for example, were considered good medicine for the brain.

23. Quoted in Webster (1979), 302.

24. Bacon (1683), 97, 107.

25. Geoghegan, (1957), 10–17.

26. Moran (2005), 31–2.

27. See Webster (1993), Pagel (1982), Debus (1966), Moran (2005).

28. Paracelsus (1656), 371.

29. Ibid., 380–2.

30. Ibid., 383.

31. Ibid., 384–5.

32. Ibid., 391–7.

33. Conrad Gesner, *The Treasure of Evonymus Conteyninge the Wonderfull hid Secretes of Nature*, trans. by Peter Morwyng (London: John Daie, 1559), 94–5; quoted in Porto (2002), 13.

34. Paracelsus, 'A book of renovation and restauration', 20, included as an appendix to Paracelsus (1661).

35. Evans (1651), unpaginated, case 95.

36. Mathew (1662), 12–14.

37. Ibid., 128–9.

38. See Duffin and Campling (2002), 61–78.

39. Fuller (1952), 615; Thornborough was seventeen when he was admitted to Magdalen College, Oxford, in 1569 (*ODNB*).

40. The National Archive, Kew, London: Prerogative Court of Canterbury, Probate Records, PROB 11/293; proved 20 June 1659, and *ODNB*.

41 See Thomas Delaune, *Angliae Metropolis, or, The Present State of London*

(London, 1690), 329–30, and Christopher Merrett, *A Short View of the Frauds, and Abuses Committed by Apothecaries* (London, 1670), 15.

42. Francis Bacon, *Novum Organum* [1620], Part 2, Aphorisms, no. 87, in Bacon (1996). On the influence of chemistry on Bacon's thought, see Rees (1975a, 1975b); see also Gaukroger (2001), 175–9, and Gregory (1938). Although Paracelsus influenced him, Bacon did not hold the German physician in high esteem; he rated the Paracelsians Peter Severinus and Joseph Duchesne more highly. See also Lemmi (1972).

43. Ben Jonson, *The Alchemist*, II, i. The play was still being performed in London in the early 1660s: see John Ward's diary, Wellcome Library, MS 6175, fol. 215.

44. Bacon, *Novum Organum*, Part 2, Aphorisms, no. 85, in Bacon (1996).

45. Ibid., no. 87.

46. Quoted in Roberts (1990), 7.

47. Roberts (1990), 2–3, 6, 10.

48. Peter Cole, 'The printer to the reader', in Culpeper (1661), unpaginated.

49. Bacon, (1638), Preface (unpaginated).

50. Bacon (1640), 170.

51. Bacon (1638) 166–7, 102.

52. Bacon (1627), Century IV.

53. Ibid.

54. Hart (1633), 6. A Scottish puritan, Hart (d.1639) had probably been educated at Edinburgh University before studying at Basel University, where Paracelsus had briefly been professor of medicine. Hart travelled extensively in Europe, receiving his MD from Basel in 1609; he eventually set up practice in Northampton. See *ODNB*.

55. Hart (1633), 7.

56. Ibid.

57. Van Helmont (1662), 753, 802.

58. Ibid., 810.

59. Ibid., 754.

60. Francis Bacon noted that, through his sparse diet, Cornaro lived 'to an extraordinary Long Life; Even of an Hundred years and better, without any Decay in his Senses; And with a constant Enjoying of his Health'. Bacon, (1638), 133–4.

61. Van Helmont (1662), 753.

62. In searching for the fountain of youth, León famously discovered Florida: see Boia (2004), 69–70.

63. Van Helmont (1662), 807–9. Van Helmont claimed to have used a quarter grain (1/600th of an ounce) of the philosopher's stone to 'project' eight ounces of quicksilver into eleven grains 'of the purest Gold': see Van Helmont (1662), chapter 111, 'Life eternal', 751–2.

64. Van Helmont (1662), 810–13. Van Helmont thought that India might offer similar healing woods.

65. Ibid., 811.

66. Porto (2002), 3–4.

67. *HP*, 'Ephemerides', Part 2 (February to May 1650), 28/1/54A–B.

68. Ibid., 'Ephemerides', 1650, Part 3 (May to October 1650): HP 28/1/61B.

69. Jennifer Speake, 'Thomas Henshaw', *ODNB*.

70. Ashmole (1652), 'Prolegomena'. In 1662, John Ward recorded in his diary that there were three versions of the philosopher's stone: vegetable, animal and angelical, of which the second 'will render men immortal and cause one to see any person on what part of ye world soever and many other conceits': Wellcome Library, MS 6175, f. 116.

71. Ashmole (1966), Vol. 2, 643.

72. Evelyn to William Wotton, 30 March 1696; in Evelyn (1955), Vol. 3, 351.

73. Petty to Boyle, 15 April 1653, in Boyle (2001), Vol. 1, 143.

74. Boyle (1999–2000), Vol. 1, cix–cxii, 3–8; quotation from p. 8.

75. 'THE ELEVENTH CHAPTER. A prognostication of what shall happen to physitians, chirugeons, apothecaries, and their dependants, and alchymists, and miners' – an appendix to *Chymical, Medicinal and Chyrugical Addresses* (1655).

76. Boyle (2001), Vol. 1, 156n, and Newman (1994), 71–2. See Van Helmont (1662), ch. lxxix.

77. Boyle (1680), 'The author's preface'. On Boyle's keen interest in medicine, see Hunter (1997), 322–61; on his chemical work, see Principe (1998).

78. This essay appears in Michael Hunter's monumental edition of Boyle's works, where it is described as 'a fragment of a significant work': Boyle (1999–2000), Vol. 9, xi, 9.5–17. My quotations are taken from the character Pyrophilus, who, Hunter writes, is 'clearly intended to represent Boyle himself': ibid., 7n.

79. Towers (1766–72), Vol. 6, 241.

80. Boyle (1680), 'The author's preface'. Notably, according to Maddison, Boyle advocated and 'undoubtedly was mainly instrumental in securing' the repeal of a statute enacted during the reign of Henry IV, which had effectively outlawed alchemy by making it illegal to attempt 'the Multiplying Gold and Silver': Maddison (1969), 176. See also Boyle (2001), Vol. 6, 288–9.

81. Boyle Papers, 'Work diary 36: Accounts of conversations with travellers and virtuosi on natural phenomena, 1685–91', entry 90: www.bbk.ac.uk/Boyle/workdiaries/WD36Clean.html (accessed 28 November 2007).

82. Le Fèvre (1664), 8–10, 59.

83. Ibid., 14–16.

84. Aubrey, Vol. 2, 182.

85. Boyle (1999b), 488–9.

86. See Raymond Crawfurd, *The Last Days of Charles II* (Oxford: Clarendon Press, 1909), 79.

87. Boyle (1999d), 467–8.

88. Ibid., 468.

89. Ibid., 466; Boyle's italics.

90. Ibid., 463–4.

91. Kaplan (1993), 65.

92. Bacon (1638), 184.

93. Farr (1980), 154–5, 159.

94. Ibid., 159; for more material on this subject, see Peter Moore, *Blood and Justice: The Seventeenth-Century Parisian Doctor who made Blood Transfusion History* (Chichester: John Wiley, 2003).

95. See Schaffer (1998), 94–105.

96. Clarke to Oldenburg, April/May 1668, in Oldenburg (1965–86), Vol. 1, 364.
97. Boyle (1999b), 494–9.
98. See William Salmon, *Pharmacopœia Bateana: Or Bate's Dispensatory* (London: printed for S. Smith and B. Walford, at the Prince's Arms in St Paul's Churchyard, 1694), 604–5.
99. Graunt (1662), 'The epistle dedicatory': Graunt asks later in this book: 'Why so many have spent their times, and estates about the Art of making Gold?' 72.
100. Appleby (1987), 35: quoted from page 3 of an appendix to a 1703 edition of Moses Stringer's *Variety of Surprising Experiments* (British Library Tracts 420/3, 420/4).
101. Roberts (1990), 5.
102. Oldenburg to Augustin Boutens, 11 November 1667, original in French, Oldenburg (1965–86), Vol. 3, 590.
103. See, for example, Letter 121, Oldenburg to Johannes Michaelis, Professor of Medicine at the University of Leipzig, Paris, 6 May 1659: Oldenburg (1965–86), Vol. 1, 241–2.
104. Oldenburg to Southwell, 29 January 1659/60: ibid., 348.
105. Southwell to Oldenburg, 20 February 1659/60: ibid., ll.355–6.
106. On Martel's identification, see Oldenburg (1965–86), Vol. 5, 484n.
107. De Martel (1670), 1179–83.
108. Dickson (1997), 66. Paston's phrase 'ye time appointed' may not indicate a particular hope for a substantially *prolonged* life.
109. Digby (1669), 'Second treatise', 148 (actually the page number is 147: mispaginated in original). On Digby, see Dobbs (1971) and Dobbs (1973).
110. See Aubrey (1898), Vol. 1, 230.
111. On 'The quintessence of snakes, adders, or vipers' and on 'Viper-wine', see French (1664), 120–1. Bacon dismissed the 'superstitious' belief 'that the Flesh, of *Serpents,* and *Harts* . . . are powerfull to the Renovation of Life; Because the one casteth his Skin, the other his Hornes'. Bacon (1638), 102–3.
112. Dobbs (1973), 143–63, 146–7; see also *ODNB*.
113. See Digby (1682), 131–2.
114. Michael Foster, 'Sir Kenelm Digby', *ODNB*.
115. Baillet (1693), 260; Lindeboom (1979), 36; Carter (1983), 113.
116. Carter (1983), 17–18, 111.
117. Hartlib to Robert Boyle, 8 or 9 May 1654, in Boyle (2001), Vol. 1, 175.
118. Hartlib to John Winthrop the Younger, 16 March 1660: Hartlib Papers, 7/7/4A.
119. Whitaker (1638), 61–2.
120. Michael Foster, 'Sir Kenelm Digby', *ODNB*. Henry Oldenburg recorded in 1657 that he was told by an Italian that Pope Urban VIII only ever ate chickens or fowl that 'had been fed and fatned [sic] by ye flesh of vipers; and yt the long Popedom was ascribed (humanitus loquendo) to ye virtue of such food'. Urban VIII (Maffeo Barberini) was pope from 1623 to 1644. Oldenburg (1965–86), Vol. 1, 134.
121. That Worsley is the author and Boyle the intended recipient of this undated letter is not certain, but it seems likely, and the letter has been included in

Boyle's most recent collection of correspondence: see Hunter et al. (2001), Vol. 1, 301–18; quotation from p. 308.

122. Mortimer (2005), 113.
123. Salisbury papers, Hatfield House, General 12/19, Vol. 22, 417–18; this would not have included the costs of his physicians' bills or those of his other caregivers. My thanks to Dr Caroline Bowden for these figures.
124. See Moses Stringer, British Library Harley MS 5931, f. 26 and Loan MS 16/2. f. 248v; see also Appleby (1987).
125. Pitt (1703), 31–2.
126. Michael Hunter, 'Robert Boyle', *ODNB*.
127. Maddison (1969), 178.
128. Smith (1676), 266.
129. Ibid., 262; 50. Smith notes that there are various insects which 'by a wonderful Metamorphosis' enjoy 'several Transformations and Renovations . . . and why some such thing as this, or at least something Analogous hereunto, may not be wrought upon man, the most perfect Creature of all the earth, I am sure no one can given an account' (264–5). Perhaps indicating the level of interest in this subject, in 1683 a translation of Roger Bacon's *Cure of Old Age* was published by Richard Brown, a licentiate of the College of Physicians.
130. Kaempfer (1727), Vol. 1, 159; Vol. 2, 534–5.
131. Bentley (1693), 10–11.
132. Ibid., 14–15.
133. Ibid., 16.
134. Ibid., 16–17.
135. Graunt (1662), 15, 18.
136. Halley (1694), 655; see also Cassedy (ed.) (1973).
137. Burnet (1697), 130–1, 146–7.

Chapter 3   The Romantic Error

1. I was once invited by an Irish publisher to contribute an article to what she told me would be 'An Encyclopaedia of the Romantic Error'. I wondered if Romanticism had been such a great mistake that you could write a whole encyclopaedia about it? It seemed – to me – a harsh judgement of an important period of modern history. Fortunately I realized my mistake before voicing these concerns: her accent had fooled me. She had meant, of course, an encyclopaedia of the Romantic *era*. But my misunderstanding gives me what seems an appropriate title to this chapter.
2. Hibbert (1980), 225.
3. See Roach (2003), 199–200.
4. Hibbert (1980), 225, 267.
5. Ibid., 226.
6. Ibid., 236.
7. Higgins (1939), ch. 23.
8. Baker (1975), 49.
9. Mlle de Lespinasse, quoted in Baker (1975), 24–5.
10. Koyré (1948), 151–2.
11. Condorcet (1795), 4.

12. Hampson (1963), 227–8.
13. Greer (1966), 126.
14. Condorcet (1795), 199–200: I have corrected the translation of Condorcet's word *misère* from 'misery' to 'extreme poverty' on the advice of Dr Caroline Warman of Jesus College, Oxford.
15. 'A conversation between Diderot and D'Alembert', in *Diderot: Rameau's Nephew and Other Works*, trans. Jacques Barzun and Ralph H. Bowen, Indianapolis: Hackett Publishing, 2001, 108.
16. Condorcet (1795), 200–1.
17. Ibid., 201–2.
18. Ibid., xiii.
19. Frazer (1933), 11.
20. See Olry and Dupont (2006), 186.
21. Hibbert (1980), 267–8.
22. Louis-Gabriel-Ambroise Bonald in *Observations sur un ouvrage posthume de Condorcet* (1795), quoted in Baker (1975), 343.
23. Denham (1730), 176.
24. Lady Sarah Cowper (1644–1720), quoted in Kugler (2001), 82.
25. Sarah Churchill (1660–1744), quoted in Thane (2000), 69.
26. Temple (1701), 112, 129–30, 151–60, 171–2, 194, 195.
27. Floud and Johnson, 283: See Wrigley et al. (1997), 614.
28. Swift (1934), 202.
29. Ibid., 203.
30. Ibid., 205.
31. Ibid., 206.
32. Ibid., 207–8.
33. McManners (1985), 63.
34. Mrs Whiteway to the Earl of Orrery, 22 November 1742, in *The Correspondence of Jonathan Swift*, Vol. 5, ed. Harold Williams (Oxford: Clarendon Press, 1965), 207.
35. Lynch (1744), 30, 47–8.
36. Ibid., 18–20, 28.
37. Mackenzie (1758), 320, 429, 445.
38. De Saingermain to Boyle, 16/26 February 1680, in Boyle (2001), Vol. 5, 185.
39. De Saingermain to Boyle, 12/22 January 1685, and 14 April 1685, ibid., Vol. 6, 96–9 and 109–11.
40. Fuller (1988), 67.
41. Ibid., 106, quoting from *Souvenirs de Charles Henri Baron de Gleichen* (Paris, 1868), pp. 122–3.
42. Ibid., 106.
43. Ibid., 135; quotation from PRO State Papers 107/6, 'Foreign ambassadors intercepted', in her translation.
44. Ibid., 225; quotation from Ernst Ahasverus Heinrich Lehndorff, *Tagebücher nach seinen Kammerherrzeit*, ed. K. E. Schmidt-Lotzen, Gotha, 1921, Vol. 1, 50.
45. Ibid., 190; quotation from Lamberg's *Memorial d'un Mondain* (1775), 84–5.
46. Quoted in Childs (1988), 85.

47. Quoted in Fuller (1988), 231; the quotation comes from a letter in French in the Zentrales Staatsarchiv, Meresburg, Akte Rep. 96, Nr. 65 D, 116r–120r. Alvensleben reported that he thought the *comte* about seventy years old – an accurate estimate, as historians believe he was born about 1706.
48. Fuller (1988), 289–90.
49. William Stukeley, *Of the Spleen* (London: printed for the author, 1723), 73.
50. Richard Mead, *A Treatise Concerning the Influence of the Sun and Moon upon Human Bodies and the Diseases Thereby Produced* (London: printed for J. Brindley, 1748), v–vi.
51. Richard Blackmore, *Treatise of Consumptions and Other Distempers Belonging to the Breast and Lungs*, 2nd edn, London: John Pemberton, 1725, xii.
52. Cheyne (1743), 1–2; Rousseau (1988), 94–5.
53. Guerrini (2000), 136.
54. Anon. (1724a), 4.
55. Cheyne (1725), 266.
56. Ibid., 4.
57. Cheyne (1740), i–ii, x–xi.
58. Cheyne to Richardson, April 1742, quoted in Shapin (2003b), 283.
59. Guerrini (2000), 127.
60. Cheyne (1740), xxiv.
61. Ibid., lviii.
62. Quoted in Guerrini (2000), 130.
63. *The Correspondence of Alexander Pope* 4.46, 4 December 1736; quoted in Guerrini (2000), 155–6.
64. Johnson (1992), 257–8 (see also his letter of 1773 to Hester). Cheyne (1742), A4v.
65. Thrale (1942), 778.
66. Ibid.
67. Thane (2000), 62.
68. Anon. (1724b), xiv–xv.
69. Ibid., 15–16.
70. Guerrini (2000), 153.
71. Cheyne (1740), 26, 23.
72. Ibid., 54–5.
73. Ibid., 62.
74. Cheyne to Samuel Richardson, 29 January 1739/40, quoted in Guerrini (2000), 173.
75. Cheyne to Richardson, December 1741, quoted in Shapin (2003b), 289.
76. James Thomson, quoted in Seward (1798), Vol. 2, 378.
77. Temple (1701), 122.
78. Haller (1801), 468.
79. Cohausen (1749), 10; see also I Kings 1: 2 for a biblical example of this practice, and Boia (2004), 117 for its recommendation by Hermaan Boerhaave.
80. See Troyansky (1982), 210.
81. A. M. Lottin, *L'almanach de la vieillesse, ou notice de tous ceux que ont vécu cent ans et plus* (Paris: A. M. Lottin, 1761).

82. Rousseau (2001), 36.

83. Napoleon Bonaparte to Tissot, 1 April 1787, quoted in Troyansky (1989), 120.

84. Gruman (1973–4).

85. Published in Latin as *Index tabularum pictarum et cælatarum qvæ longævos representant*. See Petersen and Jeune (1999): Luxdorph's source for Czartan was *Das Merckwürdige Wienn* (*Remarkable Vienna*), published anonymously in 1727. The book contained a chapter on the possibility of attaining antediluvian ages in modern time. The author cites well-known popular literature on the subject, including the work of Francis Bacon.

86. Petersen and Jeune (1999).

87. Easton (1799), xviii–xix.

88. See Troyansky (1989), especially ch. 7.

89. Buffon (1971), 150–1, 159. Translations from the French text were made for me by Susannah Wilson. On Buffon, see also Boia (2004), 85–8.

90. The English physician Edward Barry also attempted to create an equation for human life. In his *Treatise on a Consumption of the Lungs* (1727) 90–1, he suggested that the body is 'gradually destroyed' by the very circulation of the blood that supports it. And the more quickly the blood moves round the body, 'the sooner Old Age will advance, and that (*caeteris paribus*) the Number of Years, to which all Men may attain, will be in *reciprocal Ratio* to the *Velocity* of their Pulses'. Barry concluded that his calculations showed 'how absolutely necessary Temperance is to prolong Life'.

91. Buffon (1971), 148–50.

92. Quoted in Walford (1983), 10.

93. Buffon (1971), 160–71, 150.

94. Buffon, *Histoire naturelle, générale et particulière. Supplément* (Paris, 1777), Vol. 4, 405–6, cited in Rousseau (2001), 38–9.

95. Haller (1801), 467–9.

96. Quoted in Gruman (1966).

97. Priestley (1768), 5, 7–8.

98. Pepper (1911), 61–2.

99. Ibid., 61.

100. Ibid., 86–7. On this whole episode, see Jessica Riskin, *Science in the Age of Sensibility* (Chicago: Chicago University Press, 2002).

101. Porter (2003), 215–16.

102. John Abernethy, *An Enquiry into the Probability and Rationality of Mr Hunter's Theory of Life* (London, 1814), 42, 48–51, quoted in Morton (2002), 18–19.

103. Hufeland (1853), xi–xiii, 1–15.

104. Ibid., 14–15.

105. Ibid., 18–23.

106. Ibid., 25

107. Ibid., 26–9.

108. Ibid., 32–5.

109. Ibid., 70–7, 87.

110. Ibid., 59–60.

111. John-Jacques Rousseau, *Discourse Upon the Origin and Foundation of the*

Inequality Among Mankind (London: printed for R. and J. Dodsley), 1761, 118.

112. Hufeland (1853), 71: this was a popular legend, but it was only as true as the claim that Parr had lived to be 152.
113. See Paracelsus (1656), 369–407; Vaughan (1633), 157.
114. Cornaro (1727), 15.
115. Bacon (1638), 221–2.
116. Maynwaringe (1670), 152–3.
117. Ibid., 160.
118. Lindeboom (1978), 91–2. See also Carter (1983), 31.
119. See Carter (1983), 109, 112.
120. Rousseau (2006), 20.
121. Locke, *Some Thoughts Concerning Education* (1693), quoted in Spadafora (1990), 167.
122. Hartley, *Observations on Man* (1749), Vol. 2, 439, quoted in Spadafora (1990), 161.
123. See Spadafora (1990), ch. 4, 'Medicine of the mind', 135–78; and G. S. Rousseau, 'Psychology', in G.S. Rousseau and Roy Porter (eds.), *The Ferment of Knowledge: Studies in the Historiography of Eighteenth-Century Science* (Cambridge: Cambridge University Press, 1980).
124. David Hartley (1749), Vol. 1, 354.
125. Marshall (1984), 49.
126. Godwin notes that he has 'no other authority to quote for this expression than the conversation of [the Welsh moral philosopher] Dr. [Richard] Price. Upon enquiry I am happy to find it confirmed to me by Mr William Morgan, the nephew of Dr Price, who recollects to have heard it repeatedly mentioned by his uncle': Godwin (1793), Vol. 2, 392, footnote. It is possible that Price had heard Franklin's remark from Priestley, or directly from Franklin himself: all were Fellows of the Royal Society.
127. Godwin (1793), Vol. 2, 392–3.
128. Quoted in Malthus (1798), start of ch. 12.
129. Godwin (1793), Vol. 2, 393–4.
130. Ibid., 395–6.
131. Ibid., 396–7.
132. Ibid., 398.
133. Ibid., 398–402.
134. Anon., *Public Characters of 1799–1800*, London, 1799, 373.
135. Samuel Taylor Coleridge, letter to 'Caius Gracchus', 2 April 1796, in his *Collected Letters*, Vol. 1, 200, quoted in Marshall (1984), 127.
136. *Monthly Review* 28 (September, 1798), in Andrew Pyle (ed.), *Population: Contemporary Responses to Thomas Malthus* (Bristol: Thomennes Press, 1994), 11.
137. Subsequent editions of the *Essay* appeared in 1806, 1807, 1817 and 1826, swelling the book to almost four times its original size. These carried a slightly rephrased subtitle, which omitted the names of Godwin and Condorcet.
138. Malthus (1798), 127–8.
139. Ibid., 130.
140. Ibid., ch. 8, 121.

141. Ibid., ch. 9, 126.
142. Ibid.
143. Ibid., 126–7.
144. Ibid., 129.
145. Ibid., ch. 10. Wrigley points out that it is 'one of the more striking ironies of intellectual history' that Malthus published his book just when it is clear to us 'that his strictures were ceasing to be applicable'. Wrigley (2004), 18.
146. Malthus (1798), ch. 11.
147. Ibid., 150–4).
148. Ibid., 154–5.
149. Ibid., 157.
150. Mary Shelley, Introduction to the third edition of *Frankenstein* (1831), reproduced in Shelley (1969), 9.
151. Ibid. In the slightly revised text of the third edition, Shelley adds this line to Victor Frankenstein's account of his education: 'I was not unacquainted with the more obvious laws of electricity.' Morton (ed.) (2002), 172.
152. William Hazlitt, *The Spirit of the Age: Or Contemporary Portraits* (London: Henry Colburn, 2nd edn, 1825), 29.
153. 'Roger Dodsworth: The reanimated Englishman', in Shelley (1976), 43–50, first published in 1863. Also see another of her stories, 'Valerius: The reanimated Roman', in Shelley (1976), 332–44.
154. 'The mortal immortal', in Shelley (1976), 219–30; first published in 1833.

Chapter 4   From Regeneration to Degeneration

1. Winwood Reade (1838–75) was a Scottish writer and explorer of Africa. His study of the history of Western civilization, *The Martyrdom of Man*, was a fantastically popular book; in print until as recently as 1999, it went through countless editions.
2. Jefferies (1883), ch. 7, 119.
3. Ibid., ch. 8, 123.
4. Ibid., 125–6.
5. Ibid., ch. 9, 131–4.
6. Ibid., 136–7.
7. Ibid., ch. 10, 153–4.
8. Ibid., ch. 12, 176.
9. Williamson (1937), 7.
10. Matthews and Treitel (1994), 187–8.
11. Ibid., 214–15. See Jefferies's Notebook XIX, British Library MS 58820, fols 58 and 81.
12. Jefferies (1883), ch. 6, 108.
13. Williamson (1937), 30, quoting the account left by an unnamed friend of Jefferies.
14. David Hume, *The Immortality of the Soul: Two Essays* (London, 1777), quoted in de Pater (1984), 192–3.
15. William Wollaston, *The Religion of Nature Delineated* (1724), quoted in Walker (1977), 39.

16. Henry Grove, *Spectator* 626 (29 November 1714), quoted in Walker (1977), 27.

17. La Mettrie (1750), 47.

18. Quoted in Randall, *The Making of a Modern Mind: A Survey of the Intellectual Background of the Present Age* (Boston: Houghton Mifflin), 1940, 564.

19. Matthew Arnold, 'Dover Beach', in *Poems by Matthew Arnold* (London: J. M. Dent, 1948), 85–6.

20. Thrale (1942), 775–7.

21. Ibid.

22. Sinclair (1816), Appendix, 5.

23. Ibid., Appendix, 6: his wording here is clearly copied from Hufeland.

24. Ibid., Appendix, 14.

25. Lambe (1815), 189.

26. 'Aged persons', a review of Curtis's *Observations on the Preservation of Health*, in *Chamber's Edinburgh Journal* 305 (December 1837), 360.

27. Lambe (1815), 190.

28. H. Wyndham, *William Lambe MD . . . a Pioneer of Reformed Diet* (1940), 24, quoted in *ODNB*.

29. Lambe (1815), 242, 190.

30. *The Times*, 8 October 1868, quoted in Rosenberg (1998), 715.

31. See James Johnson, *A Practical Treatise on Derangements of the Liver, Digestive Organs, and Nervous System* (London: printed for the author, 2nd edn), 1818.

32. Johnson (1837), 62.

33. Ibid., 62.

34. Ibid., 62–3. Another example of the late survival of traditional conceptions is Joel Pinney's *The Duration of Human Life and its Three Eras* (London: Longman), 1856.

35. 'Suspended animation', *Chamber's Journal of Popular Literature, Science and Arts* 636 (March 1876), 158.

36. Harris (1873), 75.

37. Gardner (1874), 39–43, 150–1, 158–62.

38. Floud and Johnson (2004), 283; see Wrigley et al. (1997), 614.

39. 'Aged persons', a review of Curtis's *Observations on the Preservation of Health*, in *Chamber's Edinburgh Journal* 305 (December 1837), 360.

40. Charles Dickens, *Hard Times* (London: Bradbury & Evans), 1854, 89: quoted in Mitchell and Mitchell (2004), 36.

41. Abraham de Moivre, *Annuities on Life: Second Edition, Plainer, Fuller, and More Correct than the Former* (London: printed for the author, by Henry and George Wordfall, 1743), x.

42. See Mitchell and Mitchell (2004).

43. Ibid.

44. See P. F. Hooker, 'Benjamin Gompertz', *Journal of the Institute of Actuaries*, 91 (1965): 202–12.

45. Gompertz (1825), 517.

46. Kirkwood (2000), 33.

47. Gompertz (1825), 516–17.

48. Holland (1862), 114.

49. Flourens (1855), 54, 75.
50. Holland (1862), 105, 110, 115.
51. Ibid., 110–12.
52. Ibid., 116.
53. Van Oven (1853), xiii–xiv.
54. Ibid., appendices and notes.
55. *Bentley's Miscellany* 34 (1853), 341–2.
56. Thoms (1873), 1.
57. Ibid., 8.
58. *Notes and Queries* (3rd series), 1 (1862), 281; quoted in Thoms (1879), xiv–xv.
59. Thoms (1879).
60. Ibid., xviii–xix.
61. Thoms (1873), 266–8.
62. 'Prolongation of life', *Chamber's Journal of Popular Literature, Science and Arts* 565 (October 1874), 684.
63. Kirkwood (2000), 5, 43.
64. Quoted in Rousseau (2001), 11.
65. See Desjardins (1999).
66. Lawrence (1822), 397.
67. Jacques (1859), xii–xiv.
68. Ibid., xiv–xv.
69. Ibid., xiv–xvii.
70. Ibid., 207.
71. Ibid., 208.
72. Ibid., 215–16.
73. Ibid., 216–17.
74. Pettus (1674), 35–6.
75. Maynwaringe (1670), 9–10.
76. Ibid., 3–4.
77. Jacques (1859), 73.
78. Hufeland (1853), 180–1.
79. Ibid., 182.
80. Ibid., 183.
81. 'American advertisement', in A. Walker's *Beauty: Illustrated Chiefly by an Analysis and Classification of Beauty in Woman* (New York: J. and H. G. Langley, 1840), v.
82. Walker (1838), 364–5, 370.
83. See Daniel Pick, *Faces of Degeneration: A European Disorder, c.1848–c.1918* (Cambridge: Cambridge University Press, 1989), esp. 51–2.
84. Jacques (1859), 73.
85. Ibid., 209–10.
86. Ibid., 212, 214.
87. Fitzroy (1839), 669.
88. Darwin (1868), Vol. 1, 25.
89. Darwin (ed.) (1887), 83; Charles Darwin also relates this incident in Darwin (1868), Vol. 1, 25.
90. The question of *why* it happens (or, as Darwin put it to Huxley, 'what the devil determines each particular variation?') was a puzzling problem that

Darwin could never solve: see Freeman (1974), 213–14.
91. Darwin (ed.) (1887), letter of 1 May 1857.
92. Darwin (1989), 133–4.
93. Ibid., 138.
94. Ibid., 617–18.
95. Ibid., 619.
96. Ibid., 609.
97. Ibid., 618, 612.
98. Darwin (1887), 1.312.
99. Lankester (1870), 3.
100. Ibid., 94–5.
101. Ibid., 106.
102. Ibid., 123–4.
103. Ibid., 126–8.
104. Darwin (ed.) (1887), Charles Darwin to E. Ray Lankester, 15 March 1870, 3.120.
105. Chamberlin and Gilman (eds) (1985), vii.
106. Lankester (1880), 60–1.
107. Bowler (2003), 308–10.
108. Nordau (1895), vii, 3, 317, 536.

Chapter 5   A Brave New World

1. Private J. G. Crossley, 15th Durham Light Infantry, quoted in Martin Middlebrook (1971), 133–4.
2. Bynum (2006), 121.
3. Quoted in Shay and Wright (2000b), 72, and in Olshansky and Carnes (2001), 56–9.
4. Dormandy (1999), 199n.
5. Bynum (2006), 123–32.
6. See Pearl (1922), 23–5.
7. Metchnikoff (1910), 5–7.
8. Ibid., 85–6.
9. Ibid., 86–9.
10. Anthelme Brillat-Savarin, 'Méditation XXVI, de la mort', in *Physiologie du gout*, quoted in Illich (1976), 210–11; see Metchnikoff (1910), 126–8.
11. Metchnikoff (1921), 185–6.
12. Ibid., 195.
13. Quoted in Kirkwood (2000), 16.
14. Metchnikoff (1910), 95.
15. Ibid., 15–17.
16. Metchnikoff (1921), 171, 179, 182–3.
17. Metchnikoff (1910), 18–24.
18. Besredka (1978), 52.
19. Metchnikoff (1910), 65.
20. Chernyak and Tauber (1990), 196–7.
21. Metchnikoff (1910), 144–7.
22. Metchnikoff (1921), 197–8.
23. Ibid., Lankester's Preface, x.

24. Metchnikoff (1910), 327–8, and Metchnikoff (1921), 224.
25. Metchnikoff (1921), 232.
26. *Encyclopaedia Britannica, s. v.* 'Longevity' (Cambridge: Cambridge University Press, 1911), 977.
27. Rubinow (1913), 301.
28. 'The secret of old age', *The Times*, 20 June 1914.
29. Metchnikoff (1921), 226–7.
30. Ibid., 251.
31. Ibid., 244, 248–9, 261.
32. Ibid., 247–8.
33. Ibid., 265.
34. Besredka (1978), 83. Besredka actually writes 'Roger Bacon' rather than 'Francis' here, but he must surely be meaning the latter.
35. See http://www.yakulteurope.com/index.cfm?menuid=20, accessed 17 December 2006.
26. Pearl (1922), 43–4, 200–2: the sterile flies had a mean duration of life of 28.5 days, compared to roughly 32.2 days for Pearl's non-sterile flies.
37. *Journal of the American Medical Association*, 16 October 1920, quoted in Hamilton (1986), 41.
38. Daniel Hack Tuke, *Illustrations of the Influence of the Mind upon the Body in Health and Disease: Designed to Elucidate the Action of the Imagination* (London: J. and A. Churchill, 1872).
39. He noted that a Dr Variot had made a trial on three 'old men' aged fifty-four, fifty-six and sixty-eight, and on each the effects were 'found to be very nearly the same'. His patients knew only that they were receiving 'fortifying injections', which, Brown-Séquard argued, overcame the possibility of auto-suggestion. Two other patients to whom Variot had given injections of pure water had received 'no strengthening effect whatever'.
40. See Brown-Séquard (1889).
41. *Deutsche medizinische Wochenschrift*, quoted in Steinach and Loebel (1940), 70; see also Boia (2004), 133–4.
42. Aminoff (1993), 165.
43. Shah (2002), 433.
44. Aminoff (1993), 163.
45. Hamilton (1986), 16–17.
46. Aminoff (1993), 166–7.
47. Ibid., 167: Aminoff writes that 'it seems probable that this idea related directly to the therapeutic studies then in progress concerning testicular extracts', but notes that Murray made no reference to Brown-Séquard in his paper published in the *British Medical Journal.*
48. Aronson (2000), 506–9; Aminoff (1993), 170–1. See G. Oliver and E. Schäfer, 'On the physiological action of extract of the suprarenal capsules', *Journal of Physiology London* 16 (1894).
49. Metchnikoff (1910), 139.
50. Sengoopta (2006), 2.
51. *Journal of Experimental Medicine* (1912), 15: 516–28, quoted in Witkowski (1980), 130.
52. Witkowski (1980), 130; my translation.
53. *Scientific American*, 107 (1912): 344, 354–5. See Witkowski (1980), who

offers possible explanations for Carrel and his associate Albert Ebeling's impossible results and for the important role that the freezing of tissue cells played in subsequent versions of the experiment.
54. Bengston and Schaie (eds) (1999), 106.
55. Pearl (1922), 63.
56. *The Star* (Philadelphia), 16 January 1913, quoted in Witkowski (1980), 131.
57. A. W. Nemilow, *Leben und Tod* (Leipzig, 1927), 99, quoted in Benecke (2002), 13.
58. Witkowski (1980), 130–1, 135–6.
59. Lamont (1936), 65–6.
60. Ibid., 67.
61. Hamilton (1986), 34, 35.
62. Ibid., 26.
63. Ibid., 30.
64. Ibid., 1–2.
65. Ibid., 5–8.
66. See the *New York Times*, 1 August 1920, and Hamilton (1986).
67. Hamilton (1986), 11–12.
68. Ibid., 21.
69. Ibid., 25.
70. Ibid., 40. In fact, it is possible that castrated men actually do live longer – though the evidence remains ambiguous: see Partridge, Gems and Withers (2005), 468.
71. Following Evelyn Voronoff's untimely death in 1921, he received an annual income in excess of $300,000 from her estate. Her father, Jabez Bostwick (1830–92), had made his millions as an executive with Standard Oil – the company founded by John D. Rockefeller Sr (1839–1937), who in turn used his wealth to found the Rockefeller Institute, which employed Carrel.
72. Van Buren Dorne, '"New life" myth exploded by Dr Voronoff's wife', *New York Times*, 1 August 1920.
73. She told Van Buren Dorne that they had taken the glands from a chimpanzee because these approached man most closely, 'and the microscopic and chemical tests show that the blood of man and of the chimpanzee are identical. Nature would reject a gland differing widely from that of man. It would either die or be absorbed.' Ibid.
74. Voronoff describes the procedure in *Rejuvenation by Grafting*, trans. Fred F. Imianitoff (London: George Allen & Unwin, 1925). Hamilton explains that it was more akin to a skin graft than to modern transplant techniques: Hamilton (1986), 63.
75. 'Voronoff patient tells of new life', *New York Times*, 7 October 1922.
76. Van Buren Dorne, 'Voronoff's dramatic experiments in rejuvenation', *New York Times*, 30 May 1926. At least one of the 'after' photos actually looks quite crudely retouched.
77. 'Monkey breeding for rejuvenation; French savants hold theory is proven', *New York Times*, 24 June 1925.
78. Hamilton (1986), 66.
79. 'Voronoff predicts active life of 125 years', *New York Times*, 17 March 1927.

80. 'Voronoff foresees 150 years of life', *New York Times*, 24 May 1928.
81. 'Sees a day of supermen', *New York Times*, 31 January 1927.
82. Joseph Y. Peary, 'Rejuvenation re-evaluated', *Science*, Vol. 115, no. 2987, March 1952, 360–1.
83. *The Times*, 8 June 1928.
84. 'Condemn gland grafting', *New York Times*, 15 June 1928.
85. *The Times*, 8 June 1928.
86. *The Times*, obituary, 3 September 1951.
87. Van Buren Dorne, 'Voronoff's dramatic experiments in rejuvenation', *New York Times*, 30 May 1926.
88. 'Gland transplanting reported a failure', *New York Times*, 3 January 1928.
89. 'Discount effects of rejuvenation', *New York Times*, 11 January 1930, 3.
90. 'Dr. Voronoff, gland expert, is dead at 85', *Washington Post*, 4 September 1951.
91. Gilman (1999), 309–10.
92. Charles Willi, *Facial Rejuvenation: How to Idealise the Features and the Skin of the Face by the Latest Scientific Methods* (London: Cecil Palmer, 1926), 91, quoted in Gilman (1999), 312: see also L. J. Ludovici, *Cosmetic Scalpel: The Life of Charles Willi, Beauty-Surgeon* (Bradford-on-Avon: Moonraker, 1981).
93. For which reason I do not include any further discussion of such techniques in this book. A good recent study is Gilman (1999).
94. Quoted in Sengoopta (2006), 98.
95. Pearl (1922), 217.
96. Steinach and Loebel (1940), 149.
97. See Sengoopta (2006), 83–4.
98. Steinach and Loebel (1940), 24.
99. Benjamin (1945), 434.
100. Pearl (1922), 217–18.
101. 'Voronoff and Steinach', *Time*, 30 July 1923.
102. Dr Peter Schmidt, *Don't Be Tired: The Campaign Against Fatigue*, trans. Mary Chadwick (London: Putnam's, 1930), 24, quoted in Sengoopta (2006), 107, who explores this cultural context at much greater length.
103. Pearl (1922), 219.
104. Benjamin (1945), 433, 436. See also Wyndham (2003), 27; Sengoopta (2003), 123.
105. Kammerer (1924), 210, 216.
106. Quoted in Sengoopta (2006), 87.
107. George Sylvester Viereck, *Rejuvenation: How Steinach Makes People Young* (New York: T. Seltzer, 1923) (written under the pseudonym G. F. Corners), 1 and 95: quoted in Sengoopta (2006), 100.
108. Sengoopta (2006), 89–94.
109. Obituary, *New York Times*, 15 June 1948.
110. Wyndham (2003), 27–8.
111. Sengoopta (2006), 64.
112. Wyndham (2003), 31.
113. Haire (1924), 209–10.
114. Ibid., 9–10 and 212. Haire concluded his book with the observation: 'It is impossible to decide at present whether any of these operations actually

prolong life.' But he pointed out that there was certainly no reason to believe they shortened it.

115. Ibid., 8–9.
116. Wyndham (2003), 34
117. Quoted in Sengoopta (2005), 126.
118. Wyndham (2003), 39–40.
119. Quoted in ibid., 37.
120. Quoted in ibid., 37.
121. Quoted in ibid., 38.
122. Quoted in ibid., 38.
123. Wyndham (2003), 41.
124. Niehans's clients included Winston Churchill, Somerset Maugham and Pope Pius XII; see Trimmer (1967), 92–104.
125. 'Carrel is dubious on rejuvenation', *New York Times*, 22 December 1931.
126. Samson Wright, *Applied Physiology* (6th edn, London: Oxford University Press, 1936), 238, quoted in Sengoopta (2006), 103. On critics and the declining faith in hormones as anti-ageing therapy, see ibid., 100–10.
127. Quoted in Sengoopta (2006), 105.
128. Benjamin (1945), 437.
129. Hamilton (1986), 140–1; Joseph Y. Peary, 'Rejuvenation re-evaluated', *Science*, Vol. 115, no. 2987, March 1952, 360–1.
130. Conklin (1921), 64–5.
131. The praise is from Pearl (1922), 152.
132. Quance (1977), 9.
133. Pearl (1922), 152.
134. Quance (1977), 3.
135. 'Who shall inherit long life? On the existence of a natural process at work among human beings tending to improve the vigor and vitality of succeeding generations', *National Geographic* 35/6 (1919), 514, quoted in Quance (1977), 3.
136. See Finch et al. (2000), 14.
137. Quance (1977), 20.
138. Huxley (1926), 127.
139. *New York Times*, 29 October 1926.
140. Aldous Huxley, *Brave New World* (1932), quoted in Sengoopta (2006), 69. On Huxley's 'eternally young people', see Gilman (1999), 305–6.
141. See Sengoopta (2006), 112.
142. Conklin (1930), 587.
143. Pearl (1922), 227.
144. Schneider (1990), 274–5.
145. Carrel (1935), 318–19, quoted in Schneider (1990), 276–7.
146. Carrel (1952), 96–7, 113.
147. Ibid., 114.
148. Ibid., 150–1.
149. Carrel (1935), 290–1, quoted in Schneider (1990), 278.
150. *American Sociological Review* 1 (October 1936), 815.
151. Burleigh and Wipperman (1991), 49.
152. Dawidowicz (1977), 43–4.
153. Burleigh and Wipperman (1991), 55.

154. Glass (1989), 176.
155. Schneider (1990), 280.
156. Glass (1989), 177.
157. Carrel (1952), 179–80.
158. Boyle (1999c), 294–5 (the editors' translation from the Latin).
159. Ibid., 295.
160. John Hunter, *Lectures on the Principles of Surgery* (Philadelphia, 1841), 76.
161. Macfadyen (1899–1900), 180–2.
162. Harris (1922), 435.
163. Ibid., 437.
164. Ettinger (1965), 189.
165. Ibid., 189.
166. Ibid., xxi.
167. Ibid., 1. The French biologist Jean Rostand, who had sparked Ettinger's interest in this subject, supplied a preface to the 1965 edition of Ettinger's book in which he declared: 'we are at last even forced to concede the real possibility that the means for freezing and resuscitating human beings will one day be perfected, at however distant a time this may be' (vii).
168. Coincidentally, Evan Cooper published a book on this same idea in 1962 (*Immortality: Physically, Scientifically, Now*), under the *nom de plume* Nathan Duhring. In 1963 Cooper founded the Life Extension Foundation in Washington DC to promote freezing people.
169. 'Mechanics of freezing in living cells and tissues', *Science* 124 (1956), 515, quoted in Ettinger (1965), 10.
170. D. K. C. MacDonald, *Near Zero: The Physics of Low Temperature* (London: Heinemann, 1961), quoted in Ettinger (1965), 13.
171. Quoted in Ettinger (1965), 13.
172. See, for example, Audrey U. Smith, 'Viability of supercooled and frozen mammals', *Annals of the New York Academy of Sciences* (1959), 80: 291–300, and S. A. Goldzveig and Audrey U. Smith, 'A simple method for reanimating ice-cold rats and mice', *Journal of Physiology* (1956), 132: 406–13.
173. *New York Times*, 8 June 1981.
174. Ettinger (1965), 13–14.
175. Boyle (1999c), 294.
176. Ettinger (1965), ix.
177. Ibid., 54–5; on Wilson's work and its context, see Sengoopta (2006), 166–70.
178. Selye (1957), 276; emphasis in original.
179. Ibid., 303.
180. Hayflick (1996), 199–202.
181. See John Robins, *Healthy at 100* (2006).
182. D. E. Goldman, 'Review: American way of life?' *Science*, New Series, Vol. 145, 31 July 1964, 475–6.
183. Ettinger (1965), 3.
184. Ibid., 118–20.
185. 'Body of Californian frozen in attempt to bring back life', *New York Times*, 20 January 1967.
186. Ibid., 33–4.

187. 'Frozen dreams: A matter of life and death', *Washington Post*, 1 May 1990.
188. Appleyard (2007), 202–4.
189. Ettinger (1965), 2.

Chapter 6   This Immortal Coil

1. M. Nevinson (1926), 117.
2. Vaupel (1997), 1801.
3. Ibid., 1799.
4. Ibid., 1802; '1.2 million people will reach 100th birthday by 2074, says study': *The Guardian*, 8 February 2006, 11, citing a study by Adrian Gallop of the Government's Actuary Department.
5. See Kirkwood (1999), 33. The mortality rate continues to accelerate after the mid-eighties, but the rate of increase slows almost to zero. In mortality rate at very extreme ages (over 105), there may be a true reduction, but data are not yet sufficient to be conclusive. My thanks to Aubrey de Grey for clarifying this point.
6. Olshansky and Carnes (2001), 243.
7. See Vaupel (1997), 1802; Thatcher (1999). The Gerontology Research Group names Thomas Peters (6 April 1745–26 March 1857) of Groningen, The Netherlands, as the first supercentenarian: see the article by the Los Angeles Gerontology Research Group, 'Living and all-time world longevity record-holders over the age of 110', *Rejuvenation Research* 9 (2006), 367–8. Delina Ecker Filkins of New York State, USA (4 May 1815–4 December 1928) has also been investigated and apparently confirmed, but I have relied on the detailed work in Jeune and Vaupel (eds) (1995 and 1999), and in Vaupel (1997).
8. See Robine and Allard (1999). As this book has shown, verification of great age is difficult, and too often taken on trust or from unverifiable sources. Jeanne Calment appears to have been the only person verified to have lived beyond 120 years: the case of Shigechiyo Izumi, a Japanese man who died in 1986, supposedly aged 120, is now disputed. Sarah DeRemer Clark-Knauss (1880–1999) of the USA is, at 119, Calment's closest accepted rival. This not inconsiderable three-year gap has understandably troubled some commentators: Leonid Gavrilov and Natalia Gavrilova have pointed out that Calment's case, standing as it does so far ahead of other records, 'is a clear violation both of previous experience in record registration and of the predictions of probability theory'. But, they add, the problem lies not in her 'extreme longevity per se, since there seems to be no fixed theoretical limit to the duration of human life'. Rather, it is in 'the absence of a previous history of validated longevity records in the range of 118–121 years'. See Gavrilov and Gavrilova (2000), 403–4. Leonard Hayflick suggested in 1996 that '[m]ost gerontologists do not accept a maximum life span that exceeds about 115 years of age': Hayflick (1996), 93. For false or inaccurate claims, of which there are many, see: http://www.grg.org/Adams/G.HTM, and for a list of the oldest people ever, see http://www.grg.org/Adams/B.HTM.
9. 'Validated living supercentenarians', Gerontology Research Group (GRG) website, accessed 2 July 2007: http://www.grg.org/Adams/E.HTM. At that

date, the Japanese woman was given as the oldest living person. Her name was Yone Minagawa and she was born on 4 January 1893.

10. Ibid.
11. Ibid.
12. Oeppen and Vaupel (2002), 1029. Their graph does not take an average of individual countries, but only plots the maximum lifespan as it occurs in each record-holding country, from 1840 to 2000: these record-holding countries are Australia, Iceland, Japan, The Netherlands, New Zealand, Norway, Sweden and Switzerland. The authors write that this increase in life expectancy 'is so extraordinarily linear . . . that it may be the most remarkable regularity of mass endeavour ever observed'.
13. Kirkwood (1999), 41.
14. Westendorp (2004), 2.
15. Dr David Gems, interview with the author, London, 6 June 2007.
16. Nuland (2005), 2.
17. Dr Aubrey de Grey, interview with the author, Cambridge, 24 July 2006.
18. Unfortunately, this book (published by St Martin's Press, London) appeared too late for me to draw upon it for the present work.
19. De Grey (2004), 29–30.
20. Ibid., 38.
21. Ben Best, 'Wellness profile. Aubrey de Grey, PhD: An exclusive interview with the renowned biogerontologist', *Life Extension Magazine* (February 2006): http://www.lef.org/magazine/mag2006/feb2006_profile_01.htm, accessed 11 July 2007.
22. Rensberger (1996), 11.
23. Kirkwood (1999), 101–4.
24. Potts and Schwartz (2004), 18.
25. De Grey: personal communication, 9 July 2007.
26. Rensberger (1996), 15.
27. Ibid., 14.
28. Ibid., 12.
29. Ibid., 13–14.
30. Ibid., 12, 16–17.
31. Lee Dye, 'Lab stumbles upon "immortal" flesh: Lab workers discover skin cells that won't die', ABC News, http://www.stratatechcorp.com/about/index.php, accessed 18 April 2007.
32. Hayflick and Moorhead, 'The serial cultivation of human diploid cell strains', *Experimental Cell Research* 25 (1961), 585–621; see Hayflick (1996), 112–24 for a description of their experiment, and 127–9 for a discussion of how and why Carrel's experiment might have gone so wrong.
33. A normal human cell line will reach its limit *in vitro* in about eight months: Hayflick (1996), 130.
34. See Kirkwood (1999), 85.
35. See Rensberger (1996).
36. See Shay and Wright (2000b).
37. Hayflick (1996), 130–1; see Ewald W. Busse and Dan G. Blazer, 'The myth, history, and science of aging', in *The American Psychiatric Publishing Textbook of Geriatric Psychiatry*, 3rd edn, (Washington, DC: American Psychiatric Publishing, 2004) 3–16, at p. 5.

38. Kirkwood (1999), 98–9.
39. The cell itself may actually trigger this process (apoptosis), effectively committing suicide. Twyman (2006), 21.
40. De Grey: personal communication, 9 July 2007.
41. Shay and Wright (2000b), 75. Growth which does occur in the brain, heart and muscles is partly due to an increase in the size of these cells, not in their number, and partly due to division occurring in cells surrounding them. Hayflick (1996), 16–17.
42. Rensberger (1996), 257.
43. Hayflick (1996); Rensberger (1996), 257.
44. Hayflick (1996), xxviii.
45. Appleyard (2007), 40.
46. Kirkwood (1999), 149; see also Appleyard (2007), 38–42.
47. Hayflick (1996).
48. Rensberger (1996), 258.
49. Kirkwood (1999), 36; but Hayflick (1996), 22 states that the sea anemone is not immortal. He also argues (48–9) that one-celled animals are not immortal either, and that microbiologists who say they are (because they can continually replicate), are wrong.
50. Rensberger (1996), 264.
51. Hayflick (1996), 132.
52. Kirkwood (2006), 17; see also Partridge and Gems (2006). It appears, though, that even the very old die of *something*, and not of old age itself.
53. Hayflick (1996), 11.
54. Kirkwood (1999), 114.
55. Rensberger (1996), 261–2; Partridge and Gems (2006), 337–8.
56. Rensberger (1996), 259. Kirkwood (1999), 110–11; Shay and Wright (2000b).
57. Finch and Kirkwood (2000), 192.
58. Rensberger (1996), 260.
59. De Grey (2003), 5.
60. Rensberger (1996), 261.
61. Ibid.
62. Young (1997), 1837.
63. *Daily Express*, 16 January 2006, citing research undertaken by Prof. John Holloszy and published in the *Journal of the American College of Cardiology*.
64. Felicity Lawrence, 'Should we swallow this?', *The Guardian*, 8 February 2006.
65. Cetron and Davies (1998), 2 and 22.
66. See Finch and Kirkwood (2000), 14.
67. Kirkwood (1999), 238.
68. Hayflick (1996), 97–100.
69. Ibid., 100.
70. Ibid., 152.
71. 'How to cut obesity', *New Statesman*, 31 May 2004.
72. Sir John Krebs, Chairman of the Food Standards Agency (FSA), Great Britain, quoted in 'Official: Fat epidemic will cut life expectancy', *The Observer*, 9 November 2003.
73. See C. McCay, M. Crowell and L. Maynard, 'The effect of retarded growth

upon the length of life and upon ultimate size', *Journal of Nutrition* (1935) 10: 63–79; Walford (1983), 98–100.

74. Walford (1983), xi–xii.
57. Ibid., 2.
76. Weindruch and Walford (1988), vii.
77. Walford (1983), 112; Boia (2004), 138–40.
78. Weindruch and Walford (1988), 301–2.
79. Quoted in Appleyard (2007), 213.
80. Walford (1983), 175.
81. ww.calorierestriction.org.
82. Tristan Bettencourt, interviewed by Kara Platoni: 'Live, fast, die old', *East Bay Express*, 18 January 2006, http://eastbayexpress.com/2006-01-18/news/live-fast-die-old.
83. See T. E. Meyer, S. J. Kovacs, A. A. Ehsani, S. Klein, J. O. Holloszy, L. Fontana, 'Long-term caloric restriction ameliorates the decline in diastolic function in humans', *Journal of the American College of Cardiology* 47 (2006): 398–402. See also Partridge at al. (2005), 461, 470; 'Eat less – and live to 130', Anti-Aging Library, American Academy of Anti-Aging Medicine, www.worldhealth.net, accessed 13 May 2007.
84. Weindruch and Walford (1988), 24, 294.
85. Walker et al. (2005), 932; but see also Partridge et al. (2005).
86. Partridge and Gems (2006), 335.
87. Kenyon (2005), 457.
88. Melton (2006), 22–4.
89. Walker et al. (2005), 932.
90. Ibid., 933–4; see also Partridge and Gems (2006), 336.
91. See www.mprize.org, accessed 15 June 2007.
92. Certain species of whales live longer: see Ned Rozell, 'Bowhead whales may be the world's oldest mammals', *Alaska Science Forum*, 15 February 2001.
93. See Peter B. Medawar, *An Unsolved Problem of Biology: An Inaugural Lecture Delivered at University College, London, 6 December, 1951* (London: H. K. Lewis, 1952); George C. Williams, 'Pleiotropy, natural selection, and the evolution of senescence', *Evolution* 11 (1957), 398–411; Tom Kirkwood, 'Evolution of ageing', *Nature* 270 (24 November 1977), 301–4.
94. See Kirkwood (1999), ch. 5.
95. Ibid., p. 65.
96. Olshansky and Carnes (2001), 181.
97. Harris (2002), 70.
98. Kirkwood (1999), 69–70.
99. De Grey: personal communication, 9 July 2007.
100. See Kirkwood (1997) and Rose and Mueller (2000).
101. This is the process known as antagonistic pleiotropy: see Olshansky and Carnes (2001), 62–3.
102. See Kirkwood (1997), 1766.
103. Quoted in Olshansky and Carnes (2001), 61.
104. For a full discussion, see H. S. Kaplan and A. J. Robson, 'The emergence of humans: The coevolution of intelligence and longevity with inter-generational transfers', *Proceedings of the National Academy of Sciences of the United States of America* 99 (2002), 10221–6.

105. Kirkwood (1997), 1769–70.
106. Smith (1993), 125: in terms of a human lifespan, this change is slow; in terms of evolution, it has been described as 'rapid': see Partridge and Gems (2006), 338.
107. Fukuyama (2003), 60.
108. 'Genes and disease', National Center for Biotechnology Information; see http://www.ncbi.nlm.nih.gov, accessed 16 March 2007.
109. Fukuyama (2003), 71.
110. See Potts and Schwartz (2004).
111. Nicholas Wade, editorial comment in Aaron and Schwartz (2004), 54.
112. Ibid., 56–7.
113. Harris (2002), 66.
114. Shay and Wright (2000b), 75.
115. Shay and Wright (2000a), 23.
116. Ibid., 22.
117. Hayflick (1996), 335.
118. Ibid., 332–3, 341.
119. Ibid., 341.
120. David Gems, interview with the author, London, 6 June 2007.
121. Rose (2005), xii.
122. Boia (2004), 202.
123. Harris (2000), 59.
124. John Harris, interview with the author, Manchester, 11 January 2007.
125. De Grey (2003), 1.
126. Ibid., 3.
127. Ibid.
128. Hayflick (1996), 45.
129. De Grey (2003), 10.
130. Nuland (2005), 6.
131. EMBO Reports (November 2005), quoted in Appleyard (2007), 26–7.
132. De Grey, 'Concerns', http://www.methuselahfoundation.org/concerns. html accessed 11 July 2007.
133. See Stock (1993).
134. Ibid., 166.
135. Ibid., 168–9.
136. Ibid., 173–4.
137. I believe that, as a civilization, if we carry unthinkingly on our current course and ignore the implications to our planet of unsustainability and climate change, we are going to have much bigger issues facing us than the imminent prospect and side effects of immortality. See Jared Diamond, *Collapse: How Societies Choose to Fail or Survive* (London: Viking, 2005) and Chris Goodall, *How to Live a Low-Carbon Life: The Individual's Guide to Stopping Climate Change* (London: Earthscan, 2007) – I urge you to read these books!

Bibliography

Aaron, Henry J. and William B. Schwartz. 2004. *Coping with Methuselah: The Impact of Molecular Biology on Medicine and Society.* Washington, DC: Brookings Institution Press.

Abbott, Mary. 1996. *Life Cycles in England, 1650–1720: Cradle to Grave.* London: Routledge.

Addison, Joseph. 1721. *The Works of the Right Honourable Joseph Addison, Esq.* Vol. 3, London: Jacob Tonson.

Almond, Philip C. 1999. *Adam and Eve in Seventeenth-Century Thought.* Cambridge: Cambridge University Press.

Aminoff, Michael J. 1993. *Brown-Séquard: A Visionary of Science.* New York: Raven Press.

Anon. 1724a. *A Letter to George Cheyne, M.D., F.R.S., Shewing, The Danger of Laying Down General Rules to Those who are not Acquainted with Animal Oeconomy.* London: J. Graves.

Anon. [A Fellow of the Royal Society]. 1724b. *Remarks on Dr Cheyne's Essay of Health and Long Life. Wherein some of the Doctor's Notorious Contradictions, and False Reasonings are Laid Open.* 2nd edn, London: printed for Aaron Ward.

Appleby, John H. 1987. 'Moses Stringer (fl. 1695–1713): Iatrochemist and mineral master general'. *Ambix* 34 (1987): 31-45.

Appleyard, Bryan. 2007. *How to Live Forever or Die Trying: On the New Immortality.* London: Simon & Schuster.

Aristotle. 1910–31. *The Works of Aristotle,* translated into English under the editorship of W. D. Ross, 12 volumes. London: Oxford University Press and Oxford: Clarendon Press

Aronson, Jeffrey K. 2000. '"Where name and image meet": The argument for "adrenaline"'. *British Medical Journal* 320: 506–9.

Arrais, Edward Madeira. 1683. *Arbor Vitæ; Or, A Physical Account of the Tree of Life in the Garden of Eden.* London: printed for Tho. Fisher.

Ashmole, Elias. 1652. *Theatrum Chemicum Britannicum. Containing Severall Poeticall Pieces of our Famous English Philosophers, who have Written the Hermetique Mysteries in their Owne Ancient Language. Faithfully Collected into One Volume, with Annotations Thereon.* London: printed by J. Grismond for Nath. Brook.

Ashmole, Elias. 1658. *The Way to Bliss.* London: printed by John Grismond for Nath. Brook.

Ashmole, Elias. 1966. *Elias Ashmole (1617–1692): His Autobiographical and Historical Notes, his Correspondence, and Other Contemporary Sources Relating to his Life and Work,* ed. C. H. Josten. Oxford: Clarendon Press.

Aubrey, John. 1898. *'Brief Lives', Chiefly of Contemporaries, Set Down by John Aubrey, Between the Years 1669 and 1696,* ed. Andrew Clark. 2 vols, Oxford: Clarendon Press.

Bacon, Francis. 1627. *Sylva Sylvarum, or a Naturall Historie in Ten Centuries.* London: printed by J. H. for William Lee.

Bacon, Francis. 1638. *History Naturall and Experimentall, of Life and Death. Or of the Prolonging of Life.* London: printed by John Haviland for William Lee and Humphrey Mosley.

Bacon, Francis. 1640. *Of the Advancement and Proficience of Learning of the Partitions of Science. IX Books, translated by Gilbert Wats.* Oxford: printed by Leon. Lichfield, printer to the University, for Rob. Young and Ed. Forrest.

Bacon, Francis. 1679. *Baconiana, Or, Certaine Genuine Remains of Sr Francis Bacon.* London: printed by J. D. for Richard Chiswell.

Bacon, Francis. 1857–74 (2nd edn: 1876–83). *The Works of Francis Bacon Baron of Verulam, Viscount St Alban, and Lord High Chancellor of England.* 14 vols., ed. James Spedding, Robert L. Ellis and Douglas D. Heath. London: Longman.

Bacon, Francis. 1874. *The Letters and the Life.* In Bacon (1857–74), Vol. 14.

Bacon, Francis. 1996. *Francis Bacon: A Critical Edition of the Major Works,* ed. Brian Vickers. Oxford: Oxford University Press.

Bacon, Roger. 1683. *The Cure of Old Age, and Preservation of Youth. Translated out of the Latin, by Richard Browne, M.L. Coll. Med. Lond. Also a Physical Account of the Tree of Life, by Edw. Madeira Arrais, translated likewise out of Latin by the same hand.* London: printed for Tho. Fisher.

Baillet, Adrien. 1693. *The Life of Monsieur Des Cartes, Containing the History of his Philosophy and Works: As Also the Most Remarkable Things that Befell him During the Whole Course of his Life,* translated from the French by S. R. London: Printed for R. Simpson.

Baker, Keith Michael. 1975. *Condorcet: From Natural Philosophy to Social Mathematics.* Chicago: University of Chicago Press.

Barry, Edward. 1727. *Treatise on a Consumption of the Lungs.* London: W. and J. Innys.

Basse, William. 1619. *A Helpe to Discourse. Or, A Miscelany of Merriment Consisting of Wittie, Philosophical and Astronomicall Questions and Answers.* London: printed by Bernard Alsop for Leonard Becket.

Benecke, Mark. 2002. *The Dream of Eternal Life: Biomedicine, Aging and Immortality.* New York: Columbia University Press.

Bengston, V. L. and K. V. Schaie (eds). (1999). *Handbook of Theories of Aging.* New York: Springer.

Benjamin, Harry. 1945. 'Eugen Steinach, 1861–1944: A life of research'. *Scientific Monthly* 61: 427–42.

Bentley, Richard. 1693. *The Folly and Unreasonableness of Atheism Demonstrated from the Advantage and Pleasure of a Religious Life, The Faculties of Human Souls, The Structure of Animate Bodies, & the Origin and Frame of the World: In Eight Sermons Preached at the Lecture Founded by the Honourable Robert Boyle, Esquire.*

London: printed by J. H. for H. Mortlock.

Berzlanovich, Andrea M., Wolfgang Keil, Thomas Waldhoer, Ernst Sim, Peter Fasching and Barbara Fazeny-Dörner. 2005. 'Do centenarians die healthy? An autopsy study'. *Journals of Gerontology, Series A: Biological Sciences and Medical Sciences* 60: 862–5.

Besant, Walter. 1888. *The Eulogy of Richard Jefferies*. London: Chatto & Windus.

Besredka, Alexandre. 1978. *The Story of an Idea: E. Metchnikoff's Work: Embryogenesis, Inflammation, Immunity, Aging, Pathology, Philosophy*. Paris: Monographs of the Pasteur Institute (first published in French in 1921).

Boia, Lucian. 2004. *Forever Young: A Cultural History of Longevity*. Reaktion Books: London.

Botelho, Lynn and Pat Thane (eds). 2001. *Women and Ageing in British Society since 1500*. Harlow: Pearson Education.

Bowler, Peter J. 2003. *Evolution: The History of an Idea*. 3rd edn, Berkeley: University of California Press.

Boyd, Zacharie. 1629. *The Balme of Gilead Prepared for the Sicke*. Edinburgh: printed by John Wreittoun.

Boyle, Robert. 1680. *Experiments and Notes about the Producibleness of Chymicall Principles. Being Parts of an Appendix, design'd to be added to the Sceptical Chymist*. Oxford: printed by H. Hall for Ric. Davis.

Boyle, Robert. 1999–2000. *The Works of Robert Boyle*. Vols 1–14, ed. Michael Hunter and Edward B. Davies. London: Pickering & Chatto.

Boyle, Robert. 1999a. *General Introduction, Textual Note, Publications to 1666*. In Boyle (1999–2000), Vol. 1.

Boyle, Robert, 1999b. *Some Considerations Touching the Usefulnesse of Experimental Naturall Philosophy, Propos'd in Familiar Discourses to a Friend, by way of Invitation to the Study of it*. In Boyle (1999–2000), Vol. 3.

Boyle, Robert. 1999c. *New Experiments and Observations Touching Cold, or An Experimental History of Cold, Begun*. In Boyle (1999–2000), Vol. 4.

Boyle, Robert. 1999d. *The Origine of Formes and Qualities*. In Boyle (1999–2000), Vol. 5.

Boyle, Robert. 2001. *The Correspondence of Robert Boyle*, ed. Michael Hunter, Antonio Clericuzio and Lawrence M. Principe, London: Pickering & Chatto.

Brown, P. S. 1975. 'Medicines advertised in eighteenth-century Bath newspapers'. *Medical History* 19: 352–69.

Brown-Séquard, C. E. 1889. 'The effects produced on man by subcutaneous injections of liquid obtained from the testicles of animals'. *The Lancet* 2: 105–7.

Browne, Thomas. 1646. *Pseudodoxia Epidemic: Or, Enquiries into Very Many Received Tenents and Commonly Presumed Truths*. London: printed by T. H. for Edward Dod.

Browne, Thomas. 1669. *Hydriotaphia, Urn-Burial: Or, A Discourse of the Sepulchral Urns Lately Found in Norfolk*. London: printed for Henry Brome.

Buchan, William. 1772. *Domestic Medicine: Or A Treatise on the Prevention and Cure of Diseases by Regimen and Simple Medicines*. 2nd edn, London: W. Strahan.

Buffon, Georges-Louis LeClerc. 1971. *De l'homme*, ed. Michele Duchet. Paris: François Maspero.

Burleigh, Michael and Wolfgang Wippermann. 1991. *The Racial State: Germany, 1933–1945*. Cambridge: Cambridge University Press.

Burnet, Thomas. 1697. *The Theory of the Earth: Containing an Account of the Original of the Earth, and of all the General Changes which it Hath Already Undergone, or is to Undergo, till the Consummation of all Things.* London: printed by R. N. for Walter Kettilby.

Bynum, W. F. 2006. 'The rise of science in medicine, 1850–1913', in W. F. Bynum, Anne Hardy, Stephen Jacyna, Christopher Lawrence and E. M. Tansey (eds), *The Western Medical Tradition, 1800 to 2000.* Cambridge: Cambridge University Press.

Carnes, Bruce A., S. Jay Olshansky and Douglas Grahn. 1996. 'Continuing the search for a law of mortality'. *Population and Development Review* 22: 231–64.

Carrel, Alexis. 1935. *Man the Unknown.* London: Hamish Hamilton.

Carrel, Alexis. 1952. *Reflections on Life.* London: Hamish Hamilton.

Carter, Richard B. 1983. *Descartes' Medical Philosophy: The Organic Solution to the Mind–Body Problem.* Baltimore and London: Johns Hopkins University Press.

Cassedy, James H. (ed.). 1973. *Mortality in Pre-Industrial Times: The Contemporary Verdict.* Farnborough: Gregg International.

Cetron, Marvin and Owen Davies. 1998. *Cheating Death: The Promise and the Future Impact of Trying to Live Forever.* New York: St. Martin's Press.

Chamberlin, J. Edward and Sander L. Gilman. (eds). 1985. *Degeneration: The Dark Side of Progress.* New York: Columbia University Press.

Chernyak, Leon, and Alfred L. Tauber. 1990. 'The idea of immunity: Metchnikoff's metaphysics and science'. *Journal of the History of Biology* 23: 187–249.

Cheyne, George. 1725. *An Essay of Health and Long Life.* 2nd edn, London: printed for George Strahan and J. Leake, Bookseller at Bath.

Cheyne, George. 1740. *An Essay on Regimen. Together with Five Discourses, Medical, Moral, and Philosophical: Serving to Illustrate the Principles and Theory of Philosophical Medicin[e], and Point out Some of its Moral Consequences.* London: printed for C. Rivington, in St Paul's Churchyard; and J. Leake, Bookseller at Bath.

Cheyne, George. 1742. *The Natural Method of Cureing the Diseases of the Body, and the Disorders of the Mind Depending on the Body.* London: printed for Geo. Strahan, and John and Paul Knapton.

Cheyne, George. 1743. *Dr. Cheyne's own Account of Himself and of his Writings: Faithfully Extracted from his Various Works.* London: J. Wilford.

Childs, J. Rives. 1988. *Casanova: A New Perspective.* New York: Paragon House.

Choron, Jacques. 1963. *Death and Western Thought.* New York: Collier Books.

Cohausen, Johann Heinrich. 1749. *Hermippus Redivivus: Or, the Sage's Truth over Old Age and the Grave. Wherein a Method is laid down for Prolonging the Life and Vigour of Man. Including, a Commentary upon an Antient Inscription, in which this great Secret is Revealed; Supported by Numerous Authorities.* 2nd edn, corrected and very much enlarged, London: printed for J. Nourse.

Condorcet, Antoine-Nicolas de. 1955. *Sketch for a Historical Picture of the Progress of the Human Mind,* trans. June Barraclough with an introduction by Stuart Hampshire. London: Weidenfeld & Nicolson.

Conklin, Edwin Grant. 1921. *The Direction of Human Evolution.* London: Oxford University Press.

Conklin, Edwin Grant. 1930. 'The purposive improvement of the human race'. In

E. V. Cowdry (ed.), *Human Biology and Racial Welfare*, London: H. K. Lewis, 566–88.

Cornaro, Lewis. 1727. *Sure and Certain Methods of Attaining a Long and Healthful Life: With Means of Correcting a Bad Constitution, &c.* 4th edn, London: printed for Daniel Midwinter.

Cuffe, Henry. 1633. *The Differences of the Ages of Mans Life: Together with the Originall Causes, Progresse, and End Thereof.* London: printed by B. A. and T. F. for Lawrence Chapman.

Culpeper, Nicholas. 1661. *Pharmacopœia Londinensis: Or, The London Dispensatory.* London: Peter Cole.

Darwin, Charles. 1868. *The Variation of Animals and Plants under Domestication.* 2 vols, London: John Murray.

Darwin, Charles. 1989. *The Descent of Man, and Selection in Relation to Sex*, in *The Works of Charles Darwin*, Vols 21–2, ed. Paul H. Barrett and R. B. Freeman. London: William Pickering.

Darwin, Francis (ed.). 1887. *The Life and Letters of Charles Darwin, including an Autobiographical Chapter.* Vols 1–3, London: John Murray.

Dawidowicz, Lucy S. 1977. 'The failure of Himmler's positive eugenics', *The Hastings Center Report* 7: 43–4.

Debus, Allen G. 1966. *The English Paracelsians.* New York: Franklin Watts.

Derham, William. 1713. *Physico-Theology, or, A Demonstration of the Being and Attributes of God from his Works of Creation.* London: printed for W. Innys.

Derham, William. 1730. *Physico-Theology: Or, A Demonstration of the Being and Attributes of God, from his Works of Creation*, 10th edn. London and Dublin: printed by and for Samuel Fairbrother.

Descartes, René. 1649. *A Discourse of a Method, For the Well-Guiding of Reason, and the Discovery of Truth in the Sciences.* London: printed by Thomas Newcombe, for John Holden.

Desjardins, Bertrand. 1999. 'Validation of extreme longevity cases in the past: The French-Canadian experience', in Jeune and Vaupel (eds), 65–78.

Desjardins, Bertrand. 2001. 'Is longevity inherited? A comparison of two of the world's oldest inhabitants'. *Population: An English Selection* 13: 237–41.

Dickson, Donald R. 1997. 'Thomas Henshaw and Sir Robert Paston's pursuit of the red elixir: An early collaboration between Fellows of the Royal Society', *Notes and Records of the Royal Society of London* 51: 57–76.

Digby, Kenelm. 1669. *Of Bodies, and of Man Soul . . . With Two Discourses, Of the Powder of Sympathy, and Of the Vegetation of Plants.* London: printed by S. G. and B. G. for John Williams.

Digby, Kenelm. 1682. *A Choice Collection of Rare Secrets and Experiments in Philosophy as also Rare and Unheard-of Medicines, Menstruums and Alkahests . . . hitherto kept Secret since his Decease, but now Published for the Good and Benefit of the Publick, by George Hartman.* London: printed for the author.

Dobbs, Betty Jo. 1971. 'Studies in the natural philosophy of Sir Kenelm Digby'. *Ambix* 18: 1–25.

Dobbs, Betty Jo. 1973. 'Studies in the natural philosophy of Sir Kenelm Digby'. *Ambix* 20: 146–63.

Dobson, Mary. 1997. *Counters of Death and Disease in Early Modern England.* Cambridge: Cambridge University Press.

Dormandy, Thomas. 1999. *The White Death: A History of Tuberculosis.* London: Hambledon Press.

Duffin, Jacalyn and Barbara G. Campling. 2002. 'Therapy and disease concepts: The history (and future?) of antimony in cancer'. *Journal of the History of Medicine and Allied Sciences* 57: 61–78.

Eamon, William. 1994. *Science and the Secrets of Nature: Books of Secrets in Medieval and Early Modern Culture.* Princeton: Princeton University Press.

Easton, James. 1799. *Human Longevity: Recording the Name, Age, Place of Residence, and Year, of the Decease of 1712 Persons, who Attained a Century, & Upwards, from AD 66 to 1799, Comprising a Period of 1733 Years.* Salisbury: printed and sold by James Easton; sold also by John White, London.

Egerton, Frank N. 1966. 'The longevity of the patriarchs: A topic in the history of demography'. *Journal of the History of Ideas* 27: 575–84.

Ettinger, Robert C. 1965. *The Prospect of Immortality.* London: Sidgwick & Jackson.

Evans, John. 1651. *The Universall Medicine: Or, The Virtues of Magneticall, or Antimoniall Cup.* London: Hodgkinsonne.

Evelyn, John. 1661. *Fumifugium: Or the Inconvenience of the Aer and Smoake of London Dissipated.* London: Gabriel Bedel and Thomas Collins.

Evelyn, John. 1955. *The Diary of John Evelyn,* ed. E. S. de Beer, 6 vols. Oxford: Clarendon Press.

Faber, Albertus Otto. 1677. *De Auro Potabili Medicinali. Ad Potentissimum Principem, Carolum II.* London: printed for the author.

Farr, A. D. 1980. 'The first human blood transfusion'. *Medical History* 24: 143–62.

Finch, Caleb E. and Thomas B. L. Kirkwood. 2000. *Chance, Development and Aging.* Oxford: Oxford University Press.

FitzRoy, Robert. 1839. *Narrative of the Surveying Voyages of His Majesty's Ships Adventure and Beagle between the Years 1826 and 1836. Proceedings of the Second Expedition, 1831–36, under the Command of Captain Robert Fitz-Roy, R.N.* London: Henry Colburn.

Floud, Roderick and Paul Johnson (eds). 2004. *The Cambridge Economic History of Modern Britain, Volume 3: Industrialization, 1700—1860.* Cambridge: Cambridge University Press.

Flourens, P. 1855. *On Human Longevity and the Amount of Life upon the Globe,* translated from the French by Charles Martel. London: H. Baillière.

Frazer, James George. 1933. *Condorcet on the Progress of the Human Mind.* Oxford: Clarendon Press.

Freeman, Derek. 1974. 'The evolutionary theories of Charles Darwin and Herbert Spencer'. *Current Anthropology* 15: 211–37.

French, John. 1664. *The Art of Distillation: Or, A Treatise of the Choicest Spagyrical Preparations, Experiments, and Curiosities.* London: printed by E. Cotes for T. Williams.

Fukuyama, Francis. 2003. *Our Posthuman Future: Consequences of the Biotechnology Revolution.* London: Profile Books.

Fuller, Jean Overton. 1988. *The Comte de Saint-Germain: Last Scion of the House of Rákóczy.* London and The Hague: East–West Publications.

Fuller, Thomas. 1647. *The Historie of the Holy Warre.* London: printed by Tho. Busk.

Fuller, Thomas. 1952. *The Worthies of England*, ed. John Freeman. London: George Allen & Unwin.

Gale, Frederick M. 1971. 'Whether it is possible to prolong man's life through the use of medicine'. *Journal of the History of Medicine* 26: 391–9.

Gardner, John. 1874. *Longevity: The Means of Prolonging Life after Middle Age*. London: Henry S. King.

Gaukroger, Stephan. 2001. *Francis Bacon and the Transformation of Early-Modern Philosophy*. Cambridge: Cambridge University Press, 2001.

Gavrilov, Leonid A. and Natalia S. Gavrilova. 2000. 'Validation of exceptional longevity'. *Population and Development Review* 26: 403–4.

Gems, David. 2003. 'Is more life always better? The new biology of aging and the meaning of life'. *Hastings Center Report* 33, no. 4: 31–9.

Geoghegan, D. 1957. 'A licence to Henry VI to practise alchemy'. *Ambix* 6: 10–17.

Giglioni, Guido. 2005. 'The hidden life of matter: Techniques for prolonging life in the writings of Francis Bacon'. In Julie Robin Solomon and Catherine Gimelli Martin (eds), *Francis Bacon and the Refiguring of Early Modern Thought*, Aldershot: Ashgate, 129–44.

Gilman, Sander L. 1999. *Making the Body Beautiful: A Cultural History of Aesthetic Surgery*. Princeton: Princeton University Press.

Glass, Bentley. 1989. 'The roots of Nazi eugenics'. *Quarterly Review of Biology* 64: 175–80.

Godwin, William. 1793. *An Enquiry Concerning Political Justice, and its Influence on General Virtue and Happiness*. 2 vols, Dublin: printed for Luke White.

Gompertz, Benjamin. 1825. 'On the nature of the function expressive of the law of human mortality, and on a new mode of determining the value of life contingencies'. *Philosophical Transactions* 115: 513–83.

Grassby, Richard. 2001. *Kinship and Capitalism: Marriage, Family and Business in the English-Speaking World, 1580–1740*. Cambridge: Cambridge University Press.

Graunt, John. 1662. *Natural and Political Observations Mentioned in a Following Index, and Made Upon the Bills of Mortality. With Reference to the Government, Religion, Trade, Growth, Ayre, Diseases, and the Several Changes in the Said City*. London: Printed by Tho. Roycroft, for John Martin, James Allestry, and Tho. Dicas.

Greer, Donald. 1966. *The Incidence of the Terror during the French Revolution: A Statistical Interpretation*. Harvard, Mass.: Harvard University Press.

Gregory, Joshua C. 1938. 'Chemistry and alchemy in the natural philosophy of Sir Francis Bacon, 1561–1626'. *Ambix* 2 (1938), 93–111.

Grell, Ole Peter, and Andrew Cunningham (eds). 1993. *Medicine and the Reformation*. London: Routledge.

Grey, Aubrey de. 2003. 'An engineer's approach to the development of real anti-ageing medicine'. *Science of Aging Knowledge Environment* 1: on-line journal, www.sageke.sciencemag.org.

Grey, Aubrey de. 2004. *The War on Aging: Speculations on Some Future Chapters in the Never-Ending Story of Human Life Extension*. San Francisco: Immortality Institute, 29–45.

Griffin, Nathaniel Edward, and Hunt, Lawrence (eds). 1934. *The Farther Shore: An Anthology of World Opinion on the Immortality of the Soul*. Boston and New York: Houghton Mifflin.

Griggs, Earl Leslie (ed.). 1956–71. *The Collected Letters of Samuel Taylor Coleridge*, 6 vols. Oxford: Clarendon Press.

Gruman, Gerald. 1966. 'A history of ideas about the prolongation of life: The evolution of prolongevity hypotheses to 1800'. *Transactions of the American Philosophical Society* 56/9: 3–102.

Gruman, Gerald. 1973–4. 'Longevity'. In Philip P. Wiener (ed.), *The Dictionary of the History of Ideas: Studies of Selected Pivotal Ideas*. New York: Scribner, Vol. 3.

Guerrini, Anita. 1999. 'A diet for a sensitive soul: Vegetarianism in eighteenth-century Britain'. In Beatrice Fink (ed.), *The Cultural Topography of Food* (*Eighteenth Century Life* 23), 34–42.

Guerrini, Anita. 2000. *Obesity and Depression in the Enlightenment: The Life and Times of George Cheyne*. Norman, OK: Oklahoma University Press.

Guibbory, Achsah. 1986. *The Map of Time: Seventeenth-Century English Literature and Ideas of Pattern in History*. Urbana and Chicago: University of Illinois Press.

Guy, John. 1796. *Miscellaneous Selections: Or the Rudiments of Useful Knowledge, from the First Authorities*. 2 vols, Bristol: printed for the author by R. Edwards.

Haire, Norman. 1924. *Rejuvenation: The Work of Steinach, Voronoff, and Others*. London: George Allen & Unwin.

Haller, Albert van. 1801. *First Lines of Physiology*, translated from the 3rd Latin edn. Edinburgh: printed for Bell and Bradfute; and for Vernor and Hood, Murray and Highley, and John Cuthell; London: printed by Ad. Neill & Co.

Halley, Edmond. 1694. 'An estimate of the degrees of the mortality of mankind, drawn from curious tables of the births and funerals at the city of Breslaw; with an attempt to ascertain the price of annuities upon lives'. *Philosophical Transactions*, 596–610, 654–6.

Hamilton, David. 1986. *The Monkey Gland Affair*. London: Chatto & Windus.

Hampson . 1963. *A Social History of the French Revolution*. London: Routledge & Kegan Paul.

Harley, David. 1994. 'Political post-mortems and morbid anatomy in seventeenth-century England'. *Society for the Social History of Medicine* 7: 1–28.

Harris, D. Fraser. 1922. 'Latent life, or apparent death'. *Scientific Monthly* 14: 429–40.

Harris, George. 1873. 'The comparative longevity of animals of different species, and of man; and the probable causes which mainly conduce to promote this difference', *Journal of the Anthropological Institute of Great Britain and Ireland*, 2: 69–78.

Harris, John. 1998. *Clones, Genes, and Immortality: Ethics and the Genetic Revolution*. Oxford: Oxford University Press.

Harris, John. 2000. 'Essays on science and society: Intimations of immortality'. *Science* 288, no. 5463 (7 April): 59.

Harris, John. 2002. 'Intimations of immortality: The ethics and justice of life-extending therapies'. In *Current Legal Problems* 55, ed. M. D. A. Freeman, Oxford: Oxford University Press.

Hart, James. 1633. *Klinike, or Diet of the Diseased*. London: printed by John Beale, for Robert Allot.

Hartley, David. 1749. *Observations on Man, His Frame, His Duty, and His Expectations*. 2 vols, London: printed by S. Richardson.

Hartlib, Samuel. n.d. The Hartlib Papers. The University of Sheffield.

Hayflick, Leonard. 1996. *How and Why We Age*. New York: Ballantine.

Helmont, Jean Baptiste van. 1662. *Oriatrike, or Physick Refined. The Common Errors therein Refuted, and the Whole Art Reformed and Rectified: Being a New Rise and Progress of Phylosophy and Medicine, for the Destruction of Diseases and Prolongation of Life*. London: printed for Lodowick Loyd.

Hibbert, Christopher. 1980. *The French Revolution*. London: Allen Lane.

Higgins, E. L. 1939. *The French Revolution as Told by Contemporaries*. London: George G. Harrap.

Hobbes, Thomas. 1651. *Leviathan, or, The Matter, Forme, and Power of a Common-wealth Ecclesiastical and Civill*. London: printed for Andrew Crooke.

Hobbes, Thomas. 1994. *Thomas Hobbes: The Correspondence*, ed. Noel Malcolm. Oxford: Clarendon Press.

Hodge, M. J. S. 1990. 'Origins and species before and after Darwin', in R. C. Olby, G. N. Cantor, J. R. R. Christie and M. J. S. Hodge (eds), *Companion to the History of Modern Science*. London: Routledge.

Holland, Henry. 1862. 'Human longevity'. In *Essays on Scientific and Other Subjects, Contributed to the Edinburgh and Quarterly Review*, London: Longman, Green, Longman, Roberts & Green, 102–44.

Hufeland, Christopher William. 1853. *Hufeland's Art of Prolonging Life*, edited by Erasmus Wilson. London: John Churchill.

Hunter, Michael. 1997. 'Boyle versus the Galenists: A suppressed critique of seventeenth-century medical practice and its significance'. *Medical History* 41: 322–61.

Huxley, Julian. 1926. *Essays in Popular Science*. London: Chatto & Windus.

Hynes, Julia. 1995. 'The oldest old in pre-industrial Britain: Centenarians before 1800 – fact or fiction?' In Jeune and Vaupel (eds).

Illich, Ivan. 1976. *Limits to Medicine: Medical Nemesis: The Expropriation of Health*. London: Penguin.

Immortality Institute. 2004. *The Scientific Conquest of Death: Essays on Infinite Lifespans*. LibrosEnRed: Buenos Aires. [Available as a free pdf download from ImmInst.org.]

Jacques, Daniel Harrison. 1859. *Hints Towards Physical Perfection: or, The Philosophy of Human Beauty: Showing how to Acquire and Retain Bodily Symmetry, Health, and Vigor, Secure Long Life, and Avoid the Infirmities and Deformities of Age*. New York: Fowler & Wells.

Jardine, Lisa, and Alan Stewart. 1999. *Hostage to Fortune: The Troubled Life of Francis Bacon, 1561–1626*. London: Phoenix Giant.

Jefferies, Richard. 1883. *The Story of My Heart: My Autobiography*. London: Longmans, Green.

Jeune, Bernard and James W. Vaupel. (eds). 1995. *Exceptional Longevity: From Prehistory to the Present*. Odense: Odense University Press.

Jeune, Bernard, and James W. Vaupel. (eds). 1999. *Validation of Exceptional Longevity*. Odense: Odense University Press.

Johnson, James. 1837. *The Economy of Health, or the Stream of Human Life from the Cradle to the Grave; with Reflections Moral, Physical and Philosophical on the Successive Phases of Human Existence, the Maladies to which they are Subject, and the Dangers that may be Averted*. London: S. Highly, 2nd edn.

Johnson, Samuel. 1992. *The Letters of Samuel Johnson, Volume II: 1773–1776*, ed. Bruce Redford. Princeton: Princeton University Press.

Jonson, Ben. 1612. *The Alchemist*. London: printed by Thomas Snedham, for Walter Burre.

Kaempfer, Engelbert. 1727. *The History of Japan, Giving an Account of the Ancient and Present State and Government of that Empire*. 2 vols, London: printed for the translator, J. G. Scheuchfer.

Kammerer, Paul. 1924. *Rejuvenation and the Prolongation of Human Efficiency: Experiences with the Steinach-Operation on Man and Animals*. London: Methuen.

Kaplan, Barbara Beigun. 1993. *'Divulging of Useful Truths in Physick': The Medical Agenda of Robert Boyle*. Baltimore and London: Johns Hopkins University Press.

Kassell, Lauren. 2001. '"The Food of Angels": Simon Forman's alchemical medicine'. In Newman and Grafton (eds), 345–84.

Keill, James. 1706–7. 'An account of the death and dissection of John Bayles, of Northampton, reputed to have been 130 years old'. *Philosophical Transactions* 25: 2247–52.

Kenyon, Cynthia. 2005. 'The plasticity of aging: Insights from long-lived mutants'. *Cell* 120: 449–60.

Keynes, Geoffrey. 1966. *The Life of William Harvey*. Oxford: Clarendon Press.

Kirkman, James. 1799. *Memoirs of the Life of Charles Macklin, Esq*. London: Lackington, Allen and Co.

Kirkwood, Tom. 1997. 'The origins of human ageing'. *Philosophical Transactions: Biological Sciences*, 352/1363: 1765–72.

Kirkwood, Tom. 1999. *Time of Our Lives: Why Ageing is neither Inevitable nor Necessary*. London: Phoenix.

Kirkwood, Tom. 2006. 'Programmed for survival: Why and how do we age?' *Wellcome Focus: Ageing: Can we stop the Clock?* London: Wellcome Trust, 16–17.

Koyré, Alexander. 1948. 'Condorcet'. *Journal of the History of Ideas* 9: 131–52.

Kugler, Anne. 2001. '"I feel myself decay apace": Old age in the diary of Lady Sarah Cowper (1644–1720)'. In Botelho and Thane (eds), 66–88.

La Mettrie, Julien de. 1750. *Man a Machine*, 3rd edn. London: printed for G. Smith.

Lambe, William. 1815. *Additional Reports on the Effects of a Peculiar Regimen in Cases of Cancer, Scrofula, Consumption, Asthma, and Other Chronic Diseases*. London: J. Mawman.

Lamont, Corliss. 1936. *The Illusion of Immortality*. London: Watts.

Lankester, E. Ray. 1870. *On Comparative Longevity in Man and the Lower Animals*. London: Macmillan.

Lankester, E. Ray. 1880. *Degeneration: A Chapter in Darwinism*. London: Macmillan.

Laslett, Peter. 1991. *A Fresh Map of Life: The Emergence of the Third Age*. Cambridge, MA: Harvard University Press.

Laslett, Peter. 1999. 'The bewildering history of the history of longevity'. In Jeune and Vaupel (eds), 23–40.

Lawrence, William. 1822. *Lectures on Physiology, Zoology and the Natural History of Man*. 2nd edn, London: Benbow.

Le Febvre, Nicaise. 1664. *A Discourse upon Sir Walter Rawleigh's Great Cordial*. London: printed by J. F. for Octavian Pulleyn junior.

Lemmi, C. W. 1972. 'Mythology and alchemy in *The Wisdom of the Ancients*'. In Brian Vickers (ed.), *Essential Articles for the Study of Francis Bacon*. London:

Sidgwick & Jackson, 51–92.

Lindeboom, G. A. 1979. *Descartes and Medicine*. Amsterdam: Editions Rodopi.

Lynch, Bernard. 1744. *A Guide to Health Through the Various Stages of Life*. London: printed for the author.

Macfadyen, Allan. 1899–1900. 'On the influence of the temperature of liquid air on bacteria'. *Proceedings of the Royal Society of London* 66: 180–2.

Mackenzie, James. 1758. *The History of Health, and the Art of Preserving it: Or, An Account of all that had been recommended by Physicians and Philosophers, towards the Preservation of Health, from the most remote Antiquity to this Time*. Edinburgh: printed and sold by William Gordon.

McManners, John. 1985. *Death and the Enlightenment: Changing Attitudes to Death in Eighteenth-Century France*. Oxford: Oxford University Press.

Maddison, R. E. W. 1969. *The Life of the Honourable Robert Boyle, F.R.S.* London: Taylor & Francis.

Magalhães, João Pedro de. 2004. 'From cells to ageing: A review of models and mechanisms of cellular senescence and their impact on human ageing'. *Experimental Cell Research* 300: 1–10.

Malthus, Thomas. 1798. *Essay on the Principle of Population, as it Affects the Future Improvement of Society*. London: printed for J. Johnson.

Manton, Kenneth G., Eric Stallard and H. Dennis Tolley. 1991. 'Limits to human life expectancy: Evidence, prospects and implications'. *Population and Development Review* 17: 603–37.

Marshall, Peter H. 1984. *William Godwin*. London: Yale University Press.

Martel, Jean Pierre de. 1670. 'Of a letter written by Monsieur *de Martel* of *Montauban* to the publisher, concerning a way for the prolongation of humane life, together with some observations made in the southern parts of *France*'. *Philosophical Transactions* 58: 1179–83.

Mathew, Richard. 1662. *The Unlearned Alchymist, His Antidote. Or, A More Full and Ample Explanation of the Use, Virtue and Benefit of my PILL . . . Also, Sundry plain and easie Receits, which the Ingenuous may prepare for their own health*. London: printed for Joseph Leigh.

Matthew, H. C. G. and Brian Harrison. 2004. *Oxford Dictionary of National Biography*. Oxford: Oxford University Press.

Matthews, Hugoe, and Phyllis Treitel. 1994. *The Forward Life of Richard Jefferies: A Chronological Study*. Oxford: Petton Books.

Maynwaringe, Edward. 1670. *Vita Sana & Longa. The Preservation of Health, and Prolongation of Life. Proposed and Proved. In the due observance of Remarkable Precautions. And Daily Practicable Rules, Relating to Body and Mind, Compendiously Abstracted from the Institutions and Law of Nature*. London: printed by J. D.

Melton, Lisa. 2006. 'A family affair: Genes and ageing'. *Wellcome Focus: Ageing: Can we stop the Clock?* London: Wellcome Trust, 22–5.

Metchnikoff, Elie. 1910. *The Prolongation of Life: Optimistic Studies*, new revised edn. London: William Heinemann.

Metchnikoff, Olga. 1921. *Life of Elie Metchnikoff, 1845–1916*. London: Constable.

Middlebrook, Martin. 1971. *The First Day on the Somme, 1 July 1916*. London: Allen Lane.

Mitchell, Charlotte and Charles Mitchell. 2004. 'Wordsworth and the old men'. *Journal of Legal History*, 25: 31–52.

Moffett, Thomas. 1655. *Healths Improvement: Or, Rules Comprizing and Discovering the Nature, Method, and Manner of Preparing All Sorts of Food Used in this Nation. Corrected and Enlarged by Christopher Bennet, Doctor in Physick.* London: printed by Tho. Newcomb for Samuel Thomson.

Moivre, Abraham de. 1742. *Annuities on Lives.* 4th edn, London: printed for A. Miller.

Moran, Bruce T. 2005. *Distilling Knowledge: Alchemy, Chemistry and the Scientific Revolution.* Cambridge, MA: Harvard University Press.

Mortimer, Ian. 2005. 'The triumph of the doctors: Medical assistance to the dying, c.1570–1720'. *Transactions of the Royal Historical Society* 15: 97–116.

Morton, Timothy (ed.). 2002. *A Routledge Literary Sourcebook on Mary Shelley's Frankenstein.* London: Routledge.

Murray, John. 1830. *A Treatise on Pulmonary Consumption; Its Prevention and Remedy.* London: Whittaker, Treacher & Arnot.

Nevinson, Margaret. 1926. *Life's Fitful Fever: A Volume of Memories.* London: A. & C. Black.

Newman, William R. 1994. *Gehennical Fire: The Lives of George Starkey, an American Alchemist in the Scientific Revolution.* Cambridge, MA: Harvard University Press.

Newman, William R. and Anthony Grafton (eds). 2001. *Secrets of Nature: Astrology and Alchemy in Early Modern Europe.* London: MIT Press.

Newman, William R. and Lawrence M. Principe. 1998. 'Alchemy vs. chemistry: The etymological origins of a historiographic mistake'. *Early Science and Medicine* 3: 32–65.

Newman, William R. and Lawrence M. Principe. 2002. *Alchemy Tried in the Fire: Starkey, Boyle and the Fate of Helmontian Medicine.* Chicago: University of Chicago Press.

Nicolson, Marjorie Hope (ed.). 1930. *Conway Letters: The Correspondence of Anne, Viscountess Conway, Henry More, and their Friends, 1642–1684.* London: Oxford University Press.

Nordau, Max. 1895. *Degeneration.* London: William Heinemann.

Nuland, Sherwin. 2005. 'Do you want to live forever?' *Technology Review.* www.technologyreview.com/Biotech/14147.

Oeppen, Jim and James W. Vaupel. 2002. 'Broken limits to life expectancy'. *Science* 296: 1029–31.

Oldenburg, Henry. 1965–86. *The Correspondence of Henry Oldenburg,* ed. A. Rupert Hall and Marie Boas Hall. 13 vols, Wisconsin: University of Wisconsin Press.

Olry, Régis and Geneviève Dupont. 2006. 'Did Condorcet commit suicide?' *Journal of Medical Biography* 14: 183–6.

Olshansky, S. Jay and Bruce A. Carnes. 2001. *The Quest for Immortality: Science at the Frontiers of Aging.* New York: W.W. Norton.

Oven, Barnard van. 1853. *On the Decline of Life in Health and Disease, Being an Attempt to Investigate the Causes of Longevity, and the Best Means of Attaining a Healthful Old Age.* London: John Churchill.

Overton, Richard. 1644. *Man's Mortalitie: Or, A Treatise wherein 'tis proved, both Theologically and Philosophically, that Man (as a rationall Creature) is a Compound wholly mortall, contrary to that Common Distinction of Soule and*

Body: And that the Present Going of the Soule into Heaven or Hell is a meer Fiction. Amsterdam: printed by John Canne.

Pagel, Walter. 1982. *Paracelsus: An Introduction to Philosophical Medicine in the Era of the Renaissance.* 2nd revised edn, New York: Karger.

Paracelsus. 1656. *Paracelsus his Dispensatory and Chirurgery. The Dispensatory Contains the Choisest of his Physical remedies. And all that can be Desired of his Chirurgery . . . Faithfully Englished, by W. D.* London: printed for T. M. by Philip Chetwind. [This work includes at the end Paracelsus's *A Treatise Concerning Long Life,* 369–407.]

Paracelsus. 1661. *Paracelsus his Archidoxes: Comprised in Ten Books, Disclosing the Genuine Way of Making Quintessences, Arcanums, Magisteries, Elixirs, &c, faithfully and plainly Englished, and Published by J. H. Oxon.* London: printed for W.S.

Partridge, Linda, and David Gems. 2006. 'Beyond the evolutionary theory of ageing, from functional genomics to evo-gero'. *TRENDS in Ecology and Evolution* 21: 334–40.

Partridge, Linda, David Gems and Dominic Withers. 2005. 'Sex and death: What is the connection?' *Cell* 120: 461–72.

Pater, Wim A. de. 1984. *Immortality: Its History in the West.* Louvain: Acco.

Patrizio, Andrew and Dawn Kemp (eds). 2006. *Anatomy Acts: How We Come to Know Ourselves.* Edinburgh: Birlinn.

Pearl, Raymond. 1922. *The Biology of Death: Being a Series of Lectures Delivered at the Lowell Institute in Boston in December 1920.* Philadelphia and London: J. B. Lippincott.

Pelling, Margaret and Richard M. Smith (eds.). 1991. *Life, Death, and the Elderly: Historical Perspectives.* London: Routledge.

Pepper, William. 1911. *The Medical Side of Benjamin Franklin.* Philadelphia: William J. Campbell.

Pepys, Samuel. 1972. *The Diary of Samuel Pepys.* Vol. 6, edited by Robert Latham and William Matthews. London: G. Bell.

Pereira, Michela. 1998. '*Mater Medicinarum*: English physicians and the alchemical elixir in the fifteenth century'. In Roger French, Jon Arrizabalaga, Andrew Cunningham and Luis García-Ballester (eds), *Medicine from the Black Death to the French Disease.* Aldershot: Ashgate, 26–52.

Petersen, L.-L. B. and B. Jeune. 1999. 'Age validation of centenarians in the Luxdorph gallery'. In Jeune and Vaupel (eds).

Pettus, Sir John. 1674. *Volatiles from the History of Adam and Eve: Containing Many Unquestioned Truths, and Allowable Notions of Several Natures.* London: printed for T. Bassett.

Pitt, Robert. 1703. *The Craft and Frauds of Physick Expos'd.* London: printed for Tim. Childe.

Pope, Alexander. 1734. *The Poetical Works of Alexander Pope,* ed. Robert Carruthers, 4 vols. London: Nathaniel Cooke.

Porter, Roy. 2000. *Quacks: Fakers and Charlatans in English Medicine.* Stroud: Tempus Publishing.

Porter, Roy. 2003. *Flesh in the Age of Reason.* London: Allen Lane.

Porto, Paulo A. 2002. '"Summus atque felicissimus salium": The medical relevance of the liquor alkahest'. *Bulletin of the History of Medicine* 76: 1–29.

Potts, John T. and William B. Schwartz. 2004. 'The impact of the revolution in biomedical research on life expectancy by 2050'. In Aaron and Schwartz (eds), 16–65.

Price, Bronwen (ed.). 2002. *Francis Bacon's New Atlantis*. Manchester: Manchester University Press.

Price, Elfed Huw. 2006. 'The emergence of the doctrine of the "sentient brain" in Britain, 1650–1850'. Unpublished D.Phil thesis, Oxford University.

Priestley, J. 1768. *An Essay on the First Principles of Government and on the Nature of Political, Civil, and Religious Liberty*. Dublin.

Principe, Lawrence M. 1998. *The Aspiring Adept: Robert Boyle and his Alchemical Quest*. Princeton, NJ: Princeton University Press.

Quance, Elizabeth. 1977. 'Alexander Graham Bell, human inheritance, and the eugenics movements'. *Bulletin de Recherches* 62: unpaginated offprint: Wellcome Library, London.

Randall, John Herman. 1940. *The Making of the Modern Mind: A Survey of the Intellectual Background of the Present Age*, 2nd edn. Boston: Houghton Mifflin.

Rawley, William. 1661. *Resuscitatio, or, Bringing into Publick Light Several Pieces of the Works, Civil, Historical, Philosophical, and Theological, Hitherto Sleeping, of the Right Honourable Francis Bacon, Baron of Verulam, Viscount Saint Albans*. 2nd edn, London: printed by S. Griffin, for William Lee.

Reade, Winwoode. 1872. *The Martyrdom of Man*. London: Turner.

Rees, Graham. 1975a. 'Francis Bacon's semi-Paracelsian cosmology'. *Ambix* 22: 81–101.

Rees, Graham. 1975b. 'Francis Bacon's semi-Paracelsian cosmology and the Great Instauration'. *Ambix* 22: 161–73.

Rees, Graham. 1989. 'Medicine and medical imagery in Bacon's "Great Instauration"'. *Historical Reflections* [Canada] 16: 351–65.

Rees, Graham with Christopher Upton. 1984. *Francis Bacon's Natural Philosophy: A New Source. A Transcription of Manuscript Hardwick 72A with Translation and Commentary*. Chalfont St Giles: British Society for the History of Science, Monograph 5.

Rensberger, Boyce. 1996. *Life Itself: Exploring the Realm of the Living Cell*. Oxford: Oxford University Press.

Roach, Mary. 2003. *Stiff: The Curious Lives of Human Cadavers*. London: Penguin Books.

Roberts, Marie Mulvey. 1990. *Gothic Immortals: The Fiction of the Brotherhood of the Rosy Cross*. London: Routledge.

Roberts, Marie Mulvey. 1994. '"A physic against death": Eternal life and the Enlightenment – gender and gerontology'. In Marie Mulvey Roberts and Roy Porter (eds), *Literature and Medicine during the Eighteenth Century*. London: Routledge, 151–67.

Robine, J.-M. and M. Allard. 1999. 'Jeanne Calment: Validation of the duration of her life'. In Jeune and Vaupel (eds), 145–72.

Robinson, Tancred. 1695–7. 'A letter giving an account of one Henry Jenkins a Yorkshire man, who attained the age of 169 years, communicated by Dr Tancred Robinson, F. of the Coll. of Physitians, & R.S. with his remarks on it'. *Philosophical Transactions of the Royal Society* 19: 266–8.

Rose, Michael R. 2005. *The Long Tomorrow: How Advances in Evolutionary Biology Can Help Us to Postpone Aging*. New York: Oxford University Press.

Rose, Michael R. and Laurence D. Mueller. 2000. 'Ageing and immortality'.

Philosophical Transactions: Biological Sciences 355/1403, 1657–62.

Rosenberg, Charles E. 1998. 'Pathologies of Progress: The Idea of Civilization as Risk'. *Bulletin of the History of Medicine* 72: 714–30.

Rousseau, George S. 1988. 'Mysticism and millenarianism: "Immortal Dr Cheyne"'. In Richard Popkin, *Millenarianism and Messianism in English Literature and Thought, 1650–1800*. Leiden: Brill, 8–126.

Rousseau, George S. 2001. 'Towards a geriatric enlightenment'. In Kevin L. Cope (ed.), *1650–1850: Ideas, Aesthetics, and Inquiries in the Early Modern Era*, Vol. 6. New York: AMS Press, 3–44.

Rousseau, George S. 2006. 'A sympathy of parts: Nervous science and Scottish society'. In Patrizio and Kemp (eds), 17–30.

Rousseau, G. S. and Roy Porter (eds). 1980. *The Ferment of Knowledge: Studies in the Historiography of Eighteenth-Century Science*. Cambridge: Cambridge University Press.

Rubinow, Isaac Max. 1913. 'The old man's problem in modern industry'. In I. M. Rubinow, *Social Insurance: With Special Reference to American Conditions*. New York: Holt, 301–17; reproduced in *American Journal of Public Health* 92 (2002), 1223–6.

Ruse, Michael. 1975. 'Charles Darwin and artificial selection'. *Journal of the History of Ideas* 36: 339–50.

Salisbury Papers, Hatfield House, Hatfield, Hertfordshire (private collection).

Schäfer, Daniel. 2002. '"That senescence itself is an illness": A transitional medical concept of age and ageing in the eighteenth century'. *Medical History* 46: 525–48.

Schaffer, Simon. 1998. 'Regeneration: The body of natural philosophers in Restoration England'. In Christopher Lawrence and Steven Shapin (eds), *Science Incarnate: Historical Embodiments of Natural Knowledge*. Chicago and London: University of Chicago Press, 83–120.

Schneider, William H. 1990. *Quality and Quantity: The Quest for Biological Regeneration in Twentieth-Century France*. Cambridge: Cambridge University Press.

Selye, Hans. 1957. *The Stress of Life*. London: Longmans, Green.

Sengoopta, Chandak. 2000. 'The modern ovary: Constructions, meanings, uses'. *History of Science* 38: 425–88.

Sengoopta, Chandak. 2003. '"Dr Steinach coming to make old young!": Sex glands, vasectomy and the quest for rejuvenation in the roaring twenties'. *Endeavour* 27: 122–6.

Sengoopta, Chandak. 2006. *The Most Secret Quintessence of Life: Sex, Glands, and Hormones, 1850–1950*. Chicago and London: Chicago University Press.

Serjeantson, Richard. 2002. 'Natural knowledge in the *New Atlantis*'. In Price (ed.), 82–105.

Seward, William. 1798. *Anecdotes of Distinguished Persons*. 4th edn, 4 vols, London: printed for T. Cadell Jun. and W. Davies.

Shah, J. 2002. 'Erectile dysfunction through the ages'. *British Journal of Urology International* 90: 433–41.

Shapin, Steven. 1998. 'The philosopher and the chicken: On the dietetics of disembodied knowledge'. In Steven Shapin and Christopher Lawrence (eds), *Science Incarnate: Historical Embodiments of Natural Knowledge*, Chicago and London: Chicago University Press, 21–50.

Shapin, Steven. 2000. 'Descartes the doctor: Rationalism and its therapies'. *British Journal of the History of Science* 33: 131–54.

Shapin, Steven. 2003a. 'How to eat like a gentleman: Dietetics and ethics in early modern England'. In Charles E. Rosenberg (ed.), *Right Living: An Anglo-American Tradition of Self-Help Medicine and Hygiene.* Baltimore: Johns Hopkins University Press, 21–58.

Shapin, Steven. 2003b. 'Trusting George Cheyne: Scientific expertise, common sense, and moral authority in early eighteenth-century dietetic medicine'. *Bulletin of the History of Medicine* 77: 263–97.

Shay, Jerry W. and Woodring E. Wright 2000a. 'The use of telomerized cells for tissue engineering'. *Nature Biotechnology* 18: 22–3.

Shay, Jerry W. and Woodring E. Wright. 2000b. 'Hayflick, his limit, and cellular ageing'. *Nature Reviews: Mollecular Cell Biology* 1: 72–6.

Shelley, Mary. 1969. *Frankenstein, Or, The Modern Prometheus,* ed. James Kinsley and M. K. Joseph. Oxford: Oxford University Press.

Shelley, Mary. 1976. *Mary Shelley: Collected Tales and Stories,* ed. Charles E. Robinson. Baltimore: The Johns Hopkins University Press.

Sheskin, Arlene. 1979. *Cryonics: A Sociology of Death and Bereavement.* New York: Irvington Publishers, Inc.

Sibly, E. 1794. *The Medical Mirror; Or, Treatise on the Impregnation of the Human Female. Shewing the Origin of Diseases, and the Principles of Life and Death.* London: printed for the author.

Sinclair, John. 1816. *The Code of Health and Longevity: Or, A General View of the Rules and Principles Calculated for the Preservation of Health, and the Attainment of Long Life.* 3rd edn, London: printed for the author.

Smith, David W. E. 1993. *Human Longevity.* Oxford: Oxford University Press.

Smith, John. 1666. *The Pourtract of Old Age. Wherein is Contained a Sacred Anatomy Both of Soul, and Body, and a Perfect Account of the Infirmities of Age Incident to them Both.* 2nd edn corrected. London: printed by J. Macock, for Walter Kettilby.

Smith, Steven R. 1976. 'Growing old in seventeenth-century England'. *Albion* 8: 125–41.

Smith, Steven R. 1982. 'Growing old in an age of transition'. In Stearns (ed.), 191–208.

Sorbière, Samuel de. 1709. *A Voyage to England, Containing Many Things Relating to the State of Learning, Religion, and Other Curiosities of that Kingdom.* London: printed, and sold by J. Woodward.

Spadafora, David. 1990. *The Idea of Progress in the Eighteenth Century.* London: Yale University Press.

Spicker, Stuart F. 1990. 'Invulnerability and medicine's "promise" of immortality: Changing images of the human body during the growth of medical knowledge'. In Henk, A. M. J. Ten Have, Gerrit K. Kimsma and Stuart F. Spicker (eds), *The Growth of Medical Knowledge.* Dordrecht: Kluwer Academic Publishers, 165–74.

Starkey, George. 1657. *Natures Explication and Helmont's Vindication. Or A Short and Sure Way to a Long and Sound Life.* London: printed by E. Cotes for Thomas Alsop.

Starkey, George. 1658. *Pyrotechny Asserted and Illustrated, To be the Surest and Safest Means for Art's Triumph Over Nature's Infirmities, Being a Full and Free*

Discovery of the Medicinal Mysteries Studiously Concealed by all Artists, and only Discoverable by Fire. London: printed by R. Daniel, for Samuel Thomson.

Stearns, Peter N. (ed.). 1982. *Old Age in Preindustrial Society.* London: Holmes & Meier.

Steele, Richard. 1688. *A Discourse Concerning Old-Age, Tending to the Instruction, Caution and Comfort of Aged Persons.* London: printed by I. Astwood, for Thomas Parkhurst.

Steinach, Eugen, and Josef Loebel. 1940. *Sex and Life: Forty Years of Biological and Medical Experiments.* London: Faber and Faber.

Stock, Gregory. 1993. *Metaman: The Merging of Humans and Machines into a Global Superorganism.* London: Simon & Schuster.

Swift, Jonathan. 1934. *Gulliver's Travels and Selected Writing in Prose and Verse,* ed. John Hayward. New York: Nonesuch Press.

Talbor, Robert. 1672. *{Pyretologia}, A Rational Account of the Cause and Cure of Agues, With their Signes, Diagnostick & Prognostick.* London: printed for R. Robinson.

Taylor, John. 1635. *The Olde, Old, Very Olde Man: Or, The Age and Long Life of Thomas Parr . . .* London: printed for Henry Gosson.

Temple, William. 1701. *Miscellanea. The Third Part.* London: printed for Benjamin Tooke.

Thane, Pat. 2000. *Old Age in English History: Past Experiences, Present Issues.* Oxford: Oxford University Press.

Thatcher, A. Roger. 1999. 'Katherine Plunket: A well-documented super-centenarian in 1930'. In Jeune and Vaupel (eds), 135–44.

Thomas, Keith. 1976. 'Age and authority in early modern England'. *Proceedings of the British Academy* 62: 205–48.

Thoms, William J. 1873. *Human Longevity: Its Facts and Fictions.* London: John Murray.

Thoms, William J. 1879. *The Longevity of Man.* London: Frederic Norgate.

Thomson, George. 1666. *Loimotomia, or, The Pest Anatomized in these Following Particulars . . .* London: printed for Nath. Crouch.

Thorn, Arthur F. 1914. *Richard Jefferies and Civilisation.* London: Arthur H. Stockwell.

Thrale, Hester Lynch. 1942. *Thraliana: The Diary of Mrs Hester Lynch Thrale (Later Mrs. Piozzi), 1776–1809. Volume II: 1784–1809,* ed. Katharine C. Balderston. Oxford: Clarendon Press.

Towers, Joseph. 1766–72. *British Biography; Or, An Accurate and Impartial Account of the Lives and Writing of Eminent Persons, in Great Britain and Ireland.* 10 vols, Sherborne: printed for R. Goadby, and sold by Richard Baldwin and William Lee.

Trimmer, E. J. 1967. *Rejuvenation: The History of an Idea.* London: Robert Hale.

Troyansky, David G. 1982. 'Old age in rural enlightened Provence'. In Stearns (ed.), 209–31.

Troyansky, David G. 1989. *Old Age in the Old Regime: Image and Experience in Eighteenth-Century France.* Ithaca, NY and London: Cornell University Press.

Tryon, Thomas. 1691. *Miscellania: Or, A Collection of Necessary, Useful, and Profitable Tracts on [a] Variety of Subjects, Which for their Excellency, and Benefit of Mankind, are Compiled in One Volume.* London: T. Sowle.

Twyman, Richard. 2006. 'Wear and tear: Cells and the biology of ageing'. *Wellcome Focus: Ageing: Can we stop the Clock?* London: Wellcome Trust, 18–21.

Vaughan, R. B. 1965. 'The romantic rationalist: A study of Elie Metchnikoff '. *Medical History* 9: 201–15.

Vaughan, William. 1633. *Directions for Health, Naturall and Artificiall: Derived from the Best Physicians, as well Moderne as Antient.* 7th edn, London: printed by Thomas Harper for John Harison.

Vaupel, James W. 1997. 'The remarkable improvements in survival at older ages'. *Philosophical Transactions: Biological Sciences* 352: 1799–1804.

Voronoff, Serge. 1925. *Rejuvenation by Grafting*, translation ed. Fred F. Imianitoff. London: G. Allen & Unwin.

Wade, Nicholas. 2004. Comments appended to Potts and Schwartz, 52–7.

Walford, Roy L. 1983. *Maximum Life Span.* New York: W.W. Norton.

Walker, Alexander. 1838. *Intermarriage: Or the Mode in which, and the Causes why, Beauty, Health and Intellect, Result from Certain Unions, and Deformity, Disease and Insanity, from Others.* London: Churchill.

Walker, D. P. 1972. 'Francis Bacon and *spiritus*'. In Allen G. Debus (ed.), *Science, Medicine and Society in the Renaissance: Essays to Honor Walter Pagel*, 2 vols, London: Heinemann, 2.121–30.

Walker, Glenda, Koen Houthoof, Jacques R. Vanfleteren and David Gems. 2005. 'Dietary restriction in *C. elegans*: From rate-of-living effects to nutrient sensing pathways'. *Mechanisms of Ageing and Development* 126: 929–37.

Walker, Robert G. 1977. *Eighteenth-Century Arguments for Immortality and Johnson's 'Rasselas'.* English Literary Studies, University of Victoria, British Columbia.

Wear, Andrew. 1992. 'Health and the Environment in Early Modern England'. In Andrew Wear (ed.), *Medicine in Society: Historical Essays.* Cambridge: Cambridge University Press, 119–48.

Wear, Andrew. 2000. *Knowledge and Practice in English Medicine 1550–1680.* Cambridge: Cambridge University Press.

Webster, Charles. 1975. *The Great Instauration: Science, Medicine and Reform, 1626–1660.* London: Duckworth.

Webster, Charles. 1979. 'Alchemical and Paracelsian medicine'. In *Health, Medicine and Mortality in the Sixteenth Century.* Cambridge: Cambridge University Press.

Webster, Charles. 1993. 'Paracelsus: Medicine as popular protest'. In Ole Peter Grell and Andrew Cunningham (eds), *Medicine and the Reformation.* London: Routledge, 57–77.

Weindruch, Richard, and Roy L. Walford. 1988. *The Retardation of Aging and Disease by Dietary Restriction.* Springfield: Charles C. Thomas.

Westendorp, Rudi G. J. 2004. 'Are we becoming less disposable?' *European Molecular Biology Organization Reports* 5: 2–6.

Whitaker, Tobias. 1638. *The Tree of Humane Life, or, The Bloud of the Grape. Proving the possibilitie of maintaining Life from infancy to extreame old age without any sicknesse by the use of Wine.* London: printed by John Dawson for Henry Overton.

Williamson, Henry. 1937. *Richard Jefferies: Selections of his Work, with Details of his Life and Circumstance, his Death and Immortality.* London: Faber and Faber.

Witkowski, J. A. 1980. 'Dr Carrel's immortal cells'. *Medical History* 24: 129–42.

Wood, Anthony. 1892–4. *The Life and Times of Anthony Wood, Antiquary, of Oxford, 1632–1695, described by Himself. Collected from his Diaries and Other Papers by Andrew Clark, MA, Volumes 2 and 3.* Oxford: printed for the Oxford Historical Society.

Wrigley, E. A. 1997. *English Population History.* Cambridge: Cambridge University Press.

Wrigley, E. A. 2004. *Poverty, Progress, and Population.* Cambridge: Cambridge University Press.

Wyndham, Diana. 2003. 'Versemaking and lovemaking: W. B. Yeats' "strange second puberty": Norman Haire and the Steinach rejuvenation operation'. *Journal of the History of the Behavioral Sciences* 39: 25–50.

Young, Archie. 1997. 'Ageing and physiological functions'. *Philosophical Transactions: Biological Sciences* 352, 1837–43.

Index